Name ______________________________ Class ______________ Date ___

Skills Worksheet

Directed Reading

Section: Sexual Reproduction in Seedless Plants

Study the following steps in the life cycle of a moss. Determine the order in which the steps take place, beginning with the appearance of a new sporophyte. Write the number of each step in the space provided.

______ **1.** A zygote at the tip of a gametophyte develops into a new sporophyte, which remains attached to the gametophyte.

______ **2.** The moss spores settle to the ground, and they germinate and grow into "leafy" green gametophytes.

______ **3.** The gametophytes produce gametes inside antheridia and archegonia by mitosis.

______ **4.** When the spore capsule opens, the spores are carried away by wind or water.

______ **5.** A mature sporophyte produces spores within its spore capsule by meiosis.

______ **6.** Motile sperm swim to and fertilize eggs inside the archegonia.

Study the following steps in the life cycle of a fern. Determine the order in which the steps take place, beginning with the appearance of a new sporophyte. Write the number of each step in the space provided.

______ **7.** The zygote grows into a new sporophyte.

______ **8.** A mature sporophyte produces spores in clusters of sporangia by meiosis.

______ **9.** Sperm swim through a film of water and fertilize eggs inside archegonia.

______ **10.** The mature gametophytes produce gametes inside antheridia and archegonia by mitosis.

______ **11.** The spores fall to the ground and grow into haploid gametophytes.

Name ______________________ Class ______________ Date ____________

Skills Worksheet

Directed Reading

Section: Sexual Reproduction in Seed Plants

Complete each statement by writing the correct term or phrase in the space provided.

1. A male gametophyte of a seed plant develops into a(n)

______________ ______________ .

2. A female gametophyte develops inside a(n) ______________ , which is a multicellular structure that is part of the sporophyte.

3. The transfer of pollen grains from the male reproductive structures of a plant to the female reproductive structures of a plant is called

______________ .

4. When a pollen grain reaches a compatible female reproductive structure, a(n)

______________ ______________ emerges from the

pollen grain and grows from the pollen grain to an ovule.

Read each question, and write your answer in the space provided.

5. What is a seed coat?

____

____

6. Why are the embryos of seed plants able to survive conditions that are unfavorable for plant growth for long periods of time?

____

____

____

____

____

7. What are cotyledons?

____

____

____

8. What function do cotyledons perform for the embryo?

____

____

Study the following steps in the life cycle of a pine tree. Determine the order in which the steps take place, beginning with the appearance of a new sporophyte. Write the number of each step in the space provided.

______ **9.** The zygote and ovule develop into a seed, which grows into a new sporophyte.

______ **10.** A mature pine produces male and female cones.

______ **11.** Sperm enter the ovule through a pollen tube, and fertilization occurs.

______ **12.** Male and female spores form on the scales of the cones.

______ **13.** Pollination occurs when a pollen grain lands near an ovule.

______ **14.** Spores develop into male and female gametophytes.

In the space provided, write the letter of the description that best matches the term or phrase.

______ **15.** sepals

______ **16.** petals

______ **17.** stamen

______ **18.** anther

______ **19.** pistil

a. the part of a flower that produces ovules

b. a pollen-producing sac

c. protect a flower from damage when it is a bud

d. attract pollinators

e. a threadlike filament that is topped by an anther

In the space provided, write the letter of the term or phrase that best completes each statement or best answers each question.

______**20.** Double fertilization in the life cycle of angiosperms means that
a. a pollen grain contains two sperm cells.
b. two sperm fertilize two eggs.
c. two eggs are fertilized by one sperm.
d. one sperm fuses with the egg, and a second sperm fuses with the haploid nuclei of two other cells produced by meiosis.

______**21.** The purpose of double fertilization in the life cycle of angiosperms is to do which of the following?
a. double the number of embryos
b. produce a diploid zygote and a triploid cell that develops into the endosperm
c. decrease the number of pollen grains needed to produce one embryo
d. increase the number of cotyledons in the embryo

Name ______________________ Class ______________ Date ____________

Skills Worksheet

Directed Reading

Section: Asexual Reproduction

In the space provided, explain how the terms in each pair differ in meaning.

1. bulb, corm

____

____

____

____

2. rhizome, tuber

____

____

____

____

3. plantlet, flower

____

____

____

____

Read each question, and write your answer in the space provided.

4. What is plant propagation?

____

____

____

5. What is the difference between grafting and taking cuttings?

____

____

____

____

Name ______________________ Class ______________ Date ______________

Skills Worksheet

Active Reading

Section: Sexual Reproduction in Seedless Plants

Read the passage below. Then answer the questions that follow.

The life cycle of a fern begins when a fertilized egg, or zygote, grows into a new sporophyte. The diploid sporophyte produces spores by meiosis. The haploid spores fall to the ground and grow into haploid gametophytes. Fern gametophytes produce gametes by mitosis—eggs in archegonia and sperm in antheridia. Sperm swim to archegonia and fertilize the eggs inside them.

SKILL: READING EFFECTIVELY

Read each question, and write your answer in the space provided.

1. What is the function of meiosis in the life cycle of a fern?

____

2. What is the function of mitosis in the life cycle of a fern?

____

3. What event can occur when sperm swim to archegonia?

____

____

SKILL: SEQUENCING INFORMATION

Study the following major steps in the life cycle of a fern. Determine the order in which the steps take place, beginning with the development of a new sporophyte. Write the number of each step in the space provided.

______ **4.** Gametophytes produce gametes inside antheridia and archegonia.

______ **5.** A mature sporophyte produces spores in clusters of sporangia.

______ **6.** Sperm swim through a film of water and fertilize eggs inside archegonia.

______ **7.** A zygote develops into a new sporophyte.

______ **8.** Spores grow into a heart-shaped gametophyte.

SKILL: INTERPRETING GRAPHICS

The figure below shows the life cycle of a fern. In the space provided, write the term or phrase from the list below to identify the structure or event.

adult sporophyte	mature gametophyte	spores
archegonium	mitosis	zygote
fertilization	sporangium	

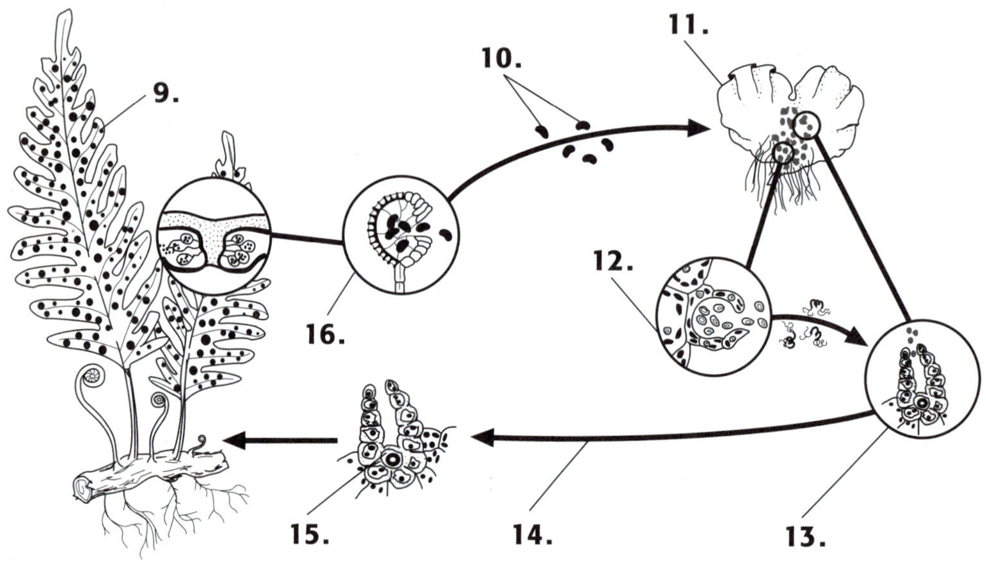

9. ______________________________

10. ______________________________

11. ______________________________

12. ______________________________

13. ______________________________

14. ______________________________

15. ______________________________

16. ______________________________

An analogy is a comparison. In the space provided, write the letter of the term that best completes the analogy.

______ **17.** Sperm is to antheridia as eggs are to

a. meiosis. **c.** archegonia.
b. mitosis. **d.** spores.

Name ________________________ Class ______________ Date ____________

Skills Worksheet

Active Reading

Section: Sexual Reproduction in Seed Plants

Read the passage below. Then answer the questions that follow.

In angiosperms, gametophytes develop within flowers. Flower parts are arranged in four concentric whorls. The outermost whorl consists of one or more **sepals,** which protect the flower from damage while it is a bud. The second whorl consists of one or more **petals,** which attract pollinators. The third whorl consists of one or more **stamens,** which produce pollen. Each stamen is made of a threadlike filament that is topped by a pollen-producing sac called an **anther.** The fourth and innermost whorl of a flower consists of one or more **pistils,** which produce ovules. Ovules develop in a pistil's swollen lower portion, which is called the **ovary.** Usually, a stalk, called the **style,** rises from the ovary. Pollen lands on and sticks to the **stigma**—the swollen, sticky tip of the style.

SKILL: READING EFFECTIVELY

Read each question, and write your answer in the space provided.

1. How are flower parts arranged?

____

2. What is the function of sepals?

____

3. What is the function of petals?

____

4. What relationship exists between stamens and the anther?

____

____

An analogy is a comparison. In the space provided, write the letter of the term that best completes the analogy.

______ **5.** Ovules are to pistils as pollen is to

a. stamens.
b. sepals.
c. stigma.
d. buds.

Name ______________________ Class ______________ Date ____________

Skills Worksheet

Active Reading

Section: Asexual Reproduction

Read the passage below. Then answer the questions that follow.

Growing new plants from seed or from vegetative parts is **plant propagation.** Plants are often propagated from structures that the plants have produced for the purpose of vegetative reproduction. The table shown below summarizes some of the methods of vegetative plant propagation that are widely used to grow plants.

Method	Description	Examples
Grafting	Small stems from one plant are attached to larger stems or roots of another plant.	Grape vines, hybrid roses, fruit trees, and nut trees
Taking cuttings	Leaves or pieces of stems or roots are cut from one plant and used to grow new individuals.	African violets, ornamental trees and shrubs, and figs
Tissue culture	Pieces of tissue from one plant are placed on a sterile medium and used to grow new individuals.	Orchids, potatoes, and many houseplants

SKILL: READING EFFECTIVELY

Read each question, and write your answer in the space provided.

1. What is plant propagation?

____

____

2. What methods of vegetative plant propagation are described in the table?

____

____

3. In addition to naming methods of vegetative plant propagation, what other information is contained in the table?

____

____

____

____

4. What occurs during grafting?

____

____

5. What types of plants are often produced by taking cuttings?

____

____

An analogy is a comparison. In the space provided, write the letter of the word or phrase that best completes the analogy.

______ **6.** Grafting is to grape vines as tissue culture is to
- **a.** figs.
- **b.** potatoes.
- **c.** fruit trees.
- **d.** hybrid roses.

Name ______________________ Class ______________ Date ____________

Skills Worksheet

Vocabulary Review

In the space provided, write the letter of the description that best matches the term or phrase.

______ **1.** a structure in seedless plants that produces eggs

______ **2.** a structure in seedless plants that produces sperm

______ **3.** a cluster of sporangia on a fern frond

______ **4.** contains a male gametophyte of a seed plant

______ **5.** the part of the sporophyte in which the female gametophyte develops

______ **6.** the transfer of pollen grains from the male to the female reproductive structure

______ **7.** grows from a pollen grain to an ovule

______ **8.** protects the embryo of a seed from mechanical injury

______ **9.** leaflike structure that is part of a plant embryo

______ **10.** one of the flower parts that protects a flower from damage when it is a bud

______ **11.** one of the flower parts that attract pollinators

______ **12.** a flower structure that consists of a threadlike filament topped with an anther

______ **13.** a pollen-producing sac

______ **14.** female reproductive part of a flower, consisting of an ovary, style, and stigma

______ **15.** a pistil's swollen lower portion

______ **16.** two sperm fusing with cells of the female gametophyte to produce both a zygote and an endosperm

a. anther
b. antheridium
c. archegonium
d. cotyledon
e. double fertilization
f. ovary
g. ovule
h. petal
i. pistil
j. pollination
k. pollen grain
l. pollen tube
m. seed coat
n. sepal
o. sorus
p. stamen

Complete each statement by writing the correct term or phrase in the space provided.

17. The type of reproduction by which plants produce offspring from vegetative parts is called ______________ ______________ .

18. Growing new plants from seeds or from vegetative parts is called ______________ ______________ .

19. In a technique called ______________ ______________ , pieces of plant tissue are placed on a sterile medium and are used to grow plants.

Name ______________________ Class ______________ Date ____________

Skills Worksheet

Science Skills

Classifying/Sequencing

Biologists have classified nonvascular plants, seedless vascular plants, and seed plants based on structural features and patterns of reproduction, many of which are listed below.

In columns 1–4, place a check mark in the boxes for the features and patterns found in each type of plant.

Feature or Pattern	**1.** Nonvascular plants	**2.** Seedless vascular plants	**3.** Gymnosperms	**4.** Angiosperms
Swimming sperm				
Pollen				
Flowers				
Independent gametophytes				
True roots and leaves				
Independent sporophytes				
Spores				
Pollination by wind				
Antheridia and archegonia				
Seeds				
Gametophytes inside male cones and female cones				
Pollination by animals				

Study the following advancements in the evolutionary history of plants. Determine the order in which the events took place. Write the number of each event in the space provided.

______ **5.** sperm swim through film of water

______ **6.** pollination

______ **7.** gametophytes produced within cones

______ **8.** vascular tissue

______ **9.** gametophytes produced within flowers

______ **10.** independent gametophytes

Name ______________________ Class ______________ Date ____________

Skills Worksheet

Concept Mapping

Using the terms and phrases provided below, complete the concept map showing the characteristics of plant reproduction.

asexual reproduction	flowers	sexual reproduction
bulbs	grafting	taking cuttings
cones	rhizomes	tissue culture
corms	runners	

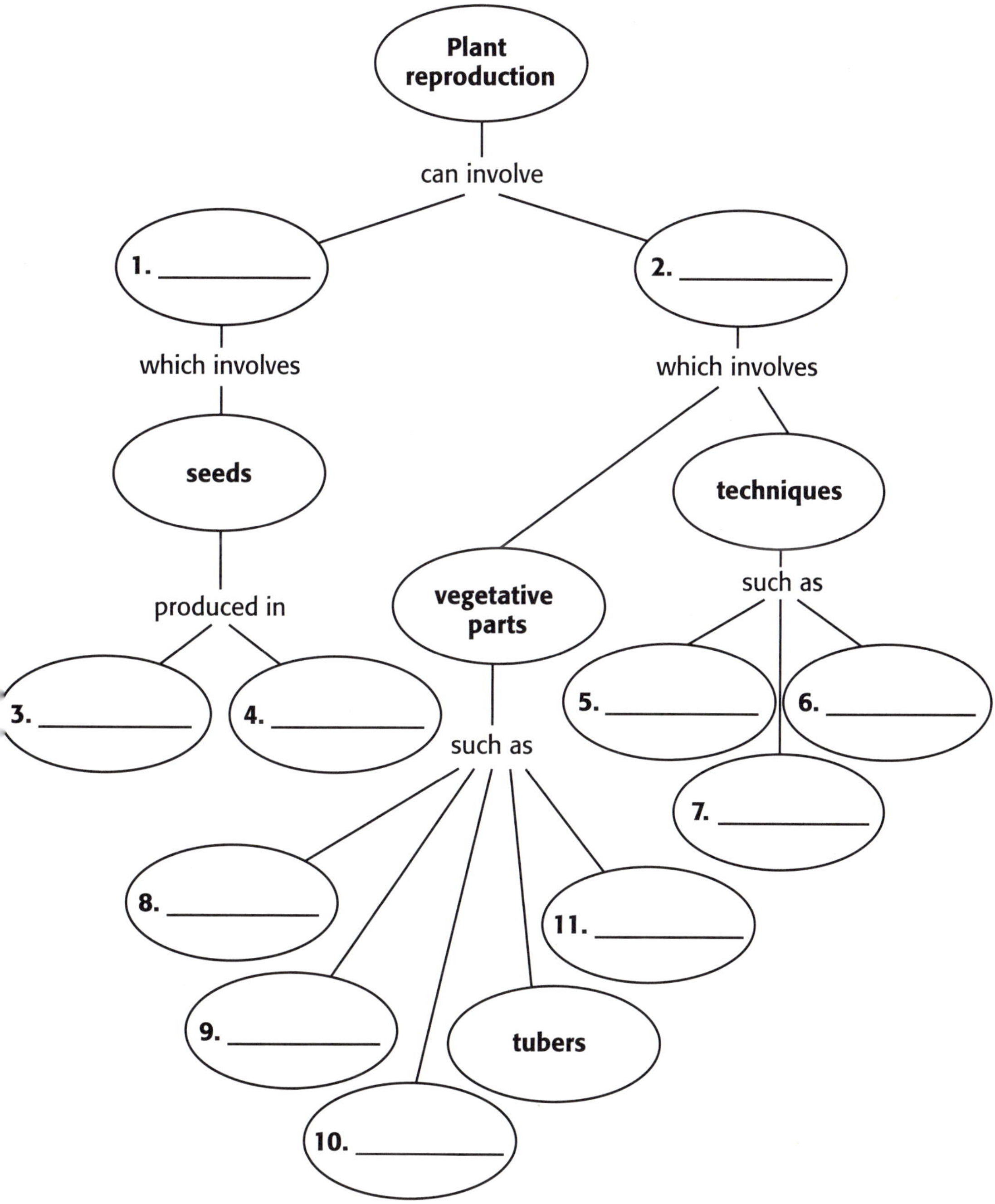

Name ______________________ Class ______________ Date ____________

Skills Worksheet

Critical Thinking

Work-Alikes

In the space provided, write the letter of the term or phrase that best describes how each numbered item functions.

_____ **1.** seed coat

_____ **2.** pine seed wing

_____ **3.** endosperm

_____ **4.** flower markings

_____ **5.** sepals

a. a helicopter blade

b. a meal

c. an airplane runway

d. a roof

e. gloves

Cause and Effect

In the space provided, write the letter of the term or phrase that best matches each cause or effect given below.

Cause	Effect
6. spore capsules open when spores are mature	______________
7. fern diploid sporophytes produce haploid spores by meiosis	______________
8. ______________	develop within tissues of gametophytes
9. ______________	they are the most successful of all plants

a. haploid spores grow into haploid gametophytes

b. they are carried away by wind or water

c. seed plants do not release their spores

d. seed plants have specialized structures within which seeds develop

Trade-offs

In the space provided, write the letter of the bad news item that best matches each numbered good news item below.

Good News

_____ **10.** Wind-pollinated flowers do not need insects for pollination.

_____ **11.** Gymnosperm cones close after pollination and protect their seeds until the seeds are mature.

_____ **12.** Seedless vascular plants thrive in moist, shady places.

Bad News

a. The process can take up to 2 years.

b. The plant has to make lots of pollen to ensure pollination.

c. They must be covered with a film of water to reproduce.

Linkages

In the spaces provided, write the letters of the two terms or phrases that are linked together by the term or phrase in the middle. The choices can be placed in any order.

13. ______ plant propagation resulting in new plants that are identical to the parent ______

14. ______ gametes fuse to form sporophyte ______

15. ______ film of water covers plant ______

16. ______ bats and moths feed on their nectar ______

a. sporophyte generation produc spores to make gametophytes

b. bulbs

c. gametes produced by gameto- phyte generation (egg and spe

d. sperm are produced by moss antheridia

e. sperm swim and fertilize egg

f. pollination occurs

g. heavily-scented, white flower open only at night

h. pieces of tissue are placed on sterile medium

Analogies

An analogy is a relationship between two pairs of terms or phrases written as a : b :: c : d. The symbol : is read as "is to," and the symbol :: is read as "as." In the space provided, write the letter of the pair of terms or phrases that best completes the analogy shown.

______ **17.** gametophytes : gametes ::
- **a.** nonvascular : antheridia
- **b.** gametes : water film
- **c.** sporophytes : spores
- **d.** liverworts : mosses

______ **18.** angiosperm : flowers ::
- **a.** gymnosperms : cones
- **b.** triploid : haploid
- **c.** angiosperms : gymnosperms
- **d.** ovules : ovary

______ **19.** dicot embryo : two cotyledons ::
- **a.** corn : shoot hook
- **b.** monocot embryo : one cotyledon
- **c.** angiosperm : embryo
- **d.** angiosperm : mature seed

______ **20.** archegonium : eggs ::
- **a.** gametophytes : spores
- **b.** sporophytes : antheridium
- **c.** sporangium : alternation
- **d.** antheridium : sperm

Name ______________________ Class ______________ Date ______________

Skills Worksheet

Test Prep Pretest

In the space provided, write the letter of the term or phrase that best completes each statement or best answers each question.

______ **1.** The gametophyte of a nonvascular plant produces sperm in a structure called a(n)
a. sporangium.
b. archegonium.
c. antheridium.
d. sorus.

______ **2.** In seedless vascular plants, the archegonia and antheridia develop on which of the following?
a. roots of the gametophytes
b. upper surfaces of the gametophytes
c. lower surfaces of the sporophytes
d. lower surfaces of the gametophytes

______ **3.** The purpose of cotyledons is to transfer nutrients to the
a. embryo.
b. leaves.
c. gametophytes.
d. endosperm.

______ **4.** In gymnosperms, the female cones produce
a. ovules.
b. pollen.
c. ovules and seeds.
d. pollen and seeds.

______ **5.** The process by which two sperm fuse with cells of the female gametophyte to produce both a zygote and endosperm is called
a. alternation of generations.
b. meiosis.
c. double fertilization.
d. asexual reproduction.

______ **6.** The male reproductive parts of a flower are called
a. petals.
b. stamens.
c. sepals.
d. pistils.

______ **7.** Kalanchoës reproduce sexually by
a. stem cuttings.
b. plantlets.
c. tiny seeds produced in flowers.
d. air roots.

______ **8.** Growing African violets from the leaf cuttings of a parent plant is one example of which of the following?
a. plant propagation
b. grafting
c. sexual reproduction
d. tissue culture

In the space provided, write the letter of the description that best matches the term or phrase.

______ **9.** pollination

______ **10.** pollen grain

______ **11.** seed coat

______ **12.** sepals

______ **13.** anther

______ **14.** pistil

______ **15.** bulb

a. the innermost whorl of a flower; produces ovules

b. forms from the hardened outer cell layers of an ovule

c. a type of modified stem with fleshy leaves

d. a pollen-producing sac at the top of a stamen

e. the outermost whorl of a flower

f. the male gametophyte of a seed plant

g. the transfer of pollen grains from a male reproductive structure of a plant to a female reproductive structure of a plant

Complete each statement by writing the correct term or phrase in the space provided.

16. Spores are produced in a(n) ______________________ in mosses. A cluster of these forms a(n) ______________________ in ferns.

17. In seed plants, male gametophytes develop into ______________________, and female gametophytes develop within the ______________________.

18. In both mosses and ferns, gametophytes produce female gametes inside ______________________ and male gametes inside ______________________.

19. In angiosperm flowers, gametophytes grow from haploid ______________________ produced by meiosis within the ______________________, which produce ovules, and the ______________________, which produce pollen.

20. Ferns, irises, and sugarcane have modified stems called ______________________, which are used in ______________________ reproduction.

21. In angiosperms, a sperm cell fuses with two other haploid cells to form a(n) ______________________ cell that develops into ______________________.

Read each question, and write your answer in the space provided.

22. How are the life cycles of a moss and a fern similar? How are they different?

____

____

____

____

____

____

____

Questions 23–25 refer to the figure below, which shows a fern life cycle.

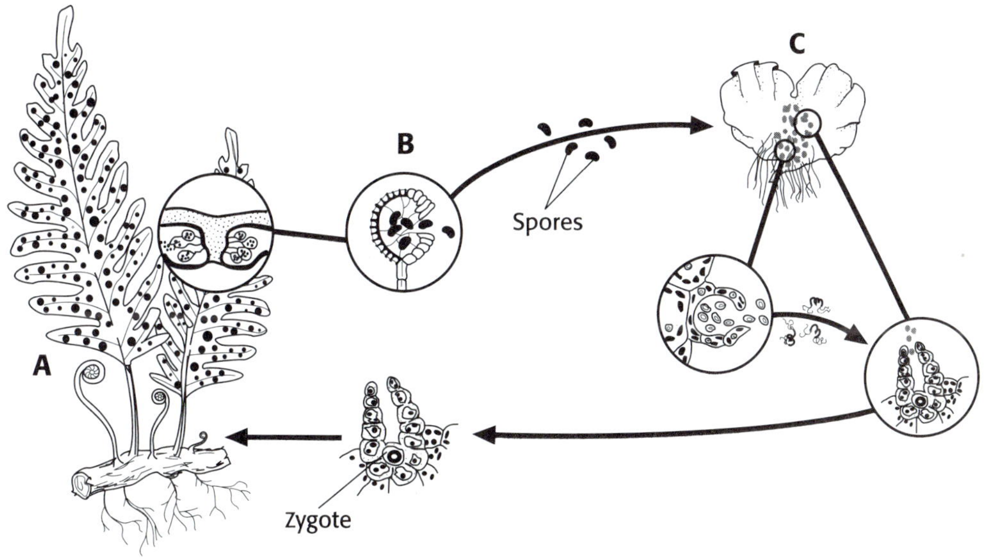

23. What generation of the fern is labeled *A*?

____

24. What is the name of the plant structure labeled *B*?

____

25. What generation of the fern is labeled *C*?

____

Name ______________________________ Class ______________ Date ____________

Assessment

Quiz

Section: Sexual Reproduction in Seedless Plants

In the space provided, write the letter of the description that best matches the term or phrase.

______ **1.** archegonium

______ **2.** antheridium

______ **3.** sorus

______ **4.** spore

______ **5.** sporophyte

a. the spore-producing generation of a seedless plant

b. produces eggs

c. produces sperm

d. carried by the wind

e. contains sporangia in ferns

In the space provided, write the letter of the term or phrase that best completes each statement or best answers each question.

______ **6.** The gametophytes of both mosses and ferns are
- **a.** found in seeds.
- **b.** haploid.
- **c.** produced on the undersides of leaves.
- **d.** diploid.

______ **7.** Compared to their gametophytes, the sporophytes of nonvascular plants are
- **a.** much smaller.
- **b.** the same size.
- **c.** much larger.
- **d.** variable in size throughout the life cycle.

______ **8.** To reproduce sexually, mosses and liverworts need
- **a.** full sunlight.
- **b.** wind.
- **c.** a film of water.
- **d.** pollination.

______ **9.** When a moss spore settles to the ground, it germinates and produces
- **a.** a "leafy" green gametophyte.
- **b.** a stalklike sporophyte.
- **c.** sperm.
- **d.** a small, green, heart-shaped gametophyte.

______ **10.** Which of the following characteristics is NOT shared by mosses and ferns?
- **a.** vascular tissue
- **b.** swimming sperm
- **c.** antheridia and archegonia
- **d.** diploid sporophyte

Name ______________________ Class ______________ Date ____________

Assessment

Quiz

Section: Sexual Reproduction in Seed Plants

In the space provided, write the letter of the description that best matches the term or phrase.

_____ **1.** sepal

_____ **2.** ovary

_____ **3.** anther

_____ **4.** cone

_____ **5.** endosperm

a. protects a flower while it is still a bud

b. produces pollen in conifers

c. location of male gametophytes in angiosperms

d. site of angiosperm fertilization

e. source of nutrition in angiosperms

In the space provided, write the letter of the term or phrase that best completes each statement or best answers each question.

_____ **6.** Compared to their sporophytes, the gametophytes of seed plants are
a. much smaller.
b. the same size.
c. much larger.
d. variable in size throughout the life cycle.

_____ **7.** A pollen grain contains
a. endosperm.
b. an embryo.
c. sperm.
d. Both (b) and (c)

_____ **8.** Inside a seed, the new sporophyte is a(n).
a. embryo.
b. cotyledon.
c. grain.
d. endosperm.

_____ **9.** In pine trees, pollen is formed
a. in male cones.
b. in female cones.
c. in tiny, inconspicuous flowers.
d. Both (a) and (b)

_____ **10.** The male parts of the flower are called
a. sepals.
b. pistils.
c. stamens.
d. ovaries.

Name ______________________ Class ______________ Date ____________

Assessment

Quiz

Section: Asexual Reproduction

In the space provided, write the letter of the description that best matches the term or phrase.

______ **1.** corm

______ **2.** runner

______ **3.** bulb

______ **4.** rhizome

______ **5.** tuber

______ **6.** cutting

a. a piece of a leaf, stem, or root

b. a horizontal, aboveground stem

c. very short stem with thick, fleshy leaves

d. very short thickened underground stem with thin, scaly leaves

e. a horizontal, underground stem

f. swollen, fleshy underground stem

In the space provided, write the letter of the term or phrase that best completes each statement or best answers each question.

______ **7.** A new African violet plant can be grown by
- **a.** planting a bulb.
- **b.** taking a cutting.
- **c.** separating a corm into pieces.
- **d.** potting a plantlet obtained from leaf edges.

______ **8.** Which of the following is NOT an example of vegetative propagation?
- **a.** seeds
- **b.** runners
- **c.** rhizomes
- **d.** tubers

______ **9.** Kalanchoës reproduce sexually by means of
- **a.** leaves.
- **b.** stems.
- **c.** roots.
- **d.** flowers.

______ **10.** Only monocots produce modified stems called
- **a.** runners.
- **b.** tubers.
- **c.** bulbs.
- **d.** rhizomes.

Name ______________________ Class ______________ Date ____________

Assessment

Chapter Test

Plant Reproduction

In the space provided, write the letters of the descriptions that best match each plant group. Some choices may be used more than once.

______ **1.** mosses	**a.** large gametophyte, small sporophyte
______ **2.** ferns	**b.** cones
______ **3.** conifers	**c.** flowers
______ **4.** angiosperms	**d.** seeds
	e. pines
	f. swimming sperm

In the space provided, write the letter of the term or phrase that best completes each statement or best answers each question.

______ **5.** Moss gametophytes grow in tightly packed clumps so that when water covers them, which of the following happens?
- **a.** Sperm can swim to nearby archegonia and fertilize the eggs.
- **b.** Eggs can float to nearby antheridia.
- **c.** The seeds can quickly germinate.
- **d.** All of the above

______ **6.** In the life cycle of a moss, the gametophyte grows from the
- **a.** egg.
- **b.** antheridium.
- **c.** zygote.
- **d.** spore.

______ **7.** Reproduction in seed plants differs from that in seedless plants in that
- **a.** a seed-plant gametophyte is larger than the sporophyte.
- **b.** a seed-plant gametophyte is part of the sporophyte.
- **c.** a seed-plant sporophyte does not release spores.
- **d.** spores are released by the gametophytes of seed plants.

______ **8.** In the life cycle of a fern, fertilization occurs within the
- **a.** antheridium.
- **b.** archegonium.
- **c.** sorus.
- **d.** sporangium.

______ **9.** Compared to their gametophytes, the sporophytes of seedless vascular plants are
- **a.** much smaller.
- **b.** the same size.
- **c.** much larger.
- **d.** variable in size throughout the life cycle.

______**10.** In seed plants, the female gametophyte is found within the
a. seed.
b. pollen tube.
c. anther.
d. ovule.

______**11.** A pollen tube delivers
a. nourishment.
b. a spore.
c. a sperm.
d. endosperm.

______**12.** The seeds of flowering plants develop from
a. ovules.
b. embryos.
c. sepals.
d. Both (a) and (b)

______**13.** After fertilization occurs, a mature pine cone opens and releases
a. ovules.
b. pollen grains.
c. seeds.
d. spores.

______**14.** Which of the following is found only in the angiosperm life cycle?
a. haploid spores
b. male and female gametophytes
c. double fertilization
d. seeds

______**15.** As a seed matures, the outer layers of the ovule harden to form the
a. seed coat.
b. endosperm.
c. cotyledons.
d. embryonic leaves.

______**16.** The female reproductive parts of a flower are called
a. petals.
b. stamens.
c. sepals.
d. pistils.

______**17.** Pine trees are pollinated by
a. wind.
b. bats.
c. bees.
d. moths.

______**18.** Which of the following flowers is likely to be pollinated by hummingbirds?
a. flowers that smell like rotten meat
b. white flowers that open at night
c. red flowers
d. tiny, pale green flowers

______**19.** Which of the following represent stems modified for vegetative reproduction?
a. runners
b. bulbs
c. tubers
d. All of the above

______**20.** A new kalanchoë plant can be grown from
a. seeds.
b. plantlets.
c. bulbs.
d. Both (a) and (b)

Name ______________________ Class ______________ Date ____________

Assessment

Chapter Test

Plant Reproduction

In the space provided, write the letter of the term or phrase that best completes each statement or best answers each question.

______ **1.** The gametophyte is much larger than the sporophyte in
a. conifers. **c.** horsetails.
b. ferns. **d.** liverworts.

______ **2.** Mosses and ferns are alike in that they can reproduce sexually only when
a. a film of water covers the gametophyte.
b. a film of water covers the sporophyte.
c. the gametophytes germinate.
d. the sporophytes are small.

______ **3.** The seed part that protects the embryo from mechanical injury and keeps out water and oxygen is called the
a. sorus. **c.** seed coat.
b. cotyledon. **d.** ovule.

______ **4.** In gymnosperms, gametophytes form on
a. flowers. **c.** stamens.
b. the scales of cones. **d.** the scales of sepals.

______ **5.** In the life cycle of a pine tree,
a. both male and female cones are produced.
b. seeds with wings travel away from the parent tree.
c. microscopic gametophytes develop from spores.
d. All of the above

______ **6.** A zygote and endosperm are produced in the process called
a. alternation of generations. **c.** double fertilization.
b. meiosis. **d.** vegetative propagation.

______ **7.** Meiosis occurs
a. as a zygote becomes a seed.
b. only within the seed coat.
c. as sporophytes reproduce.
d. whenever plant cells divide.

______ **8.** The tiny seeds produced by kalanchoë flowers result from
a. asexual reproduction. **c.** sexual reproduction.
b. vegetative reproduction. **d.** tissue culture.

______ **9.** Which of the following is NOT a method of vegetative plant propagation?
a. grafting
b. seeds
c. cuttings
d. tissue culture

Questions 10–13 refer to the figure at right, which shows the structure of a flower.

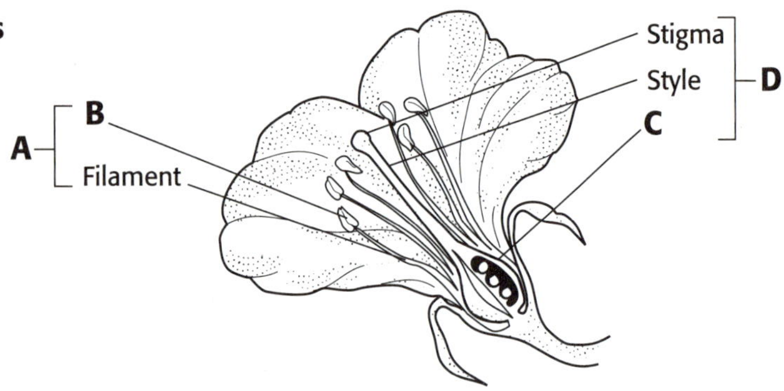

______**10.** The structure labeled B produces
a. male gametophytes.
b. seeds.
c. female gametophytes.
d. antheridia.

______**11.** Which of the following occurs during pollination?
a. Pollen grains from *B* are transferred to *C*.
b. Eggs from *C* are transferred to *B*.
c. Eggs from *B* are transferred to *D*.
d. Pollen grains from *A* are transferred to *D*.

______**12.** The pollen-producing sac is labeled
a. *A*.
b. *B*.
c. *C*.
d. *D*.

______**13.** In angiosperms, ovules are produced in the structure labeled
a. *A*.
b. *B*.
c. *C*.
d. *D*.

Complete each statement by writing the correct term or phrase in the space provided.

14. In a flower, female ______________________ develop in ovules within the ovary, which is the lower portion of the ______________________ .

15. In seed plants, sperm that develop within ______________________ grains travel through a(n) ______________________ ______________________ to the egg, which develops within a multicellular structure called a(n) ______________________ .

16. A leaflike structure of a seed that transfers nutrients to the embryo is called a(n) ______________________ .

17. Most angiosperm flowers have four basic types of parts, which are the ______________________ , the ______________________ , the ______________________ , and the ______________________ .

18. In angiosperm reproduction, the zygote and the tissues of the ovule develop into a(n) ______________________ , which grows into a new ______________________ .

19. Flowers that lack either stamens or pistils are called ______________________ flowers.

20. Plant propagation is the process of growing new plants from ______________________ or ______________________ ______________________ .

Read each question, and write your answer in the space provided.

21. Beginning with the sporophyte stage, describe the life cycle of a fern.

____

____

____

____

____

22. Describe how sperm reach eggs in the life cycle of a conifer.

____

____

____

____

____

23. In angiosperms, what happens after a pollen tube is formed?

____

____

____

____

____

24. List five examples of structures by which plants reproduce asexually.

____

____

____

____

____

25. Describe how plantlets are involved in the reproduction of kalanchoës.

____

____

____

____

____

Name ____________________ Class ____________ Date ____________

Quick Lab

DATASHEET FOR IN-TEXT LAB

Observing a Fern Gametophyte

You can observe the archegonia and antheridia of a fern gametophyte with a microscope.

MATERIALS

- prepared slide of a fern gametophyte with archegonia and antheridia
- compound microscope

Procedure

1. Examine a slide of a fern gametophyte under low power of a microscope. Move the slide until you can see a cluster of archegonia. Now, switch to high power, and focus on one archegonium. Draw and label what you see. Use a separate sheet of paper.

2. Switch back to low power, and move the slide until you can see several egg-shaped structures. These are antheridia. Now, switch to high power, and focus on one antheridium. Draw and label what you see.

Analysis

1. Describe the appearance of an archegonium and an antheridium.

____

____

____

____

2. Critical Thinking
Drawing Conclusions In which structure, an archegonium or antheridium, does the growth of a new sporophyte begin? Explain.

____

____

____

____

____

Name ______________________________ Class ______________ Date ____________

Quick Lab

DATASHEET FOR IN-TEXT LAB

Observing the Gametophytes of Pines

You can observe the gametophytes of a pine with a microscope.

MATERIALS

- prepared slides of the following: male pine cone, female pine cone, pine ovule
- hand lens
- compound microscope

Procedure

1. Examine prepared slides of male and female pine cones first with a hand lens and then under the low power of a microscope.

2. Make a sketch of each type of pine cone, and label the structures that you recognize.

3. Examine a prepared slide of a pine ovule under the low power of a compound microscope. Compare what you see with the drawing.

4. Draw a pine ovule, and label the following structures: scale, ovule, egg, pollen tube (if visible).

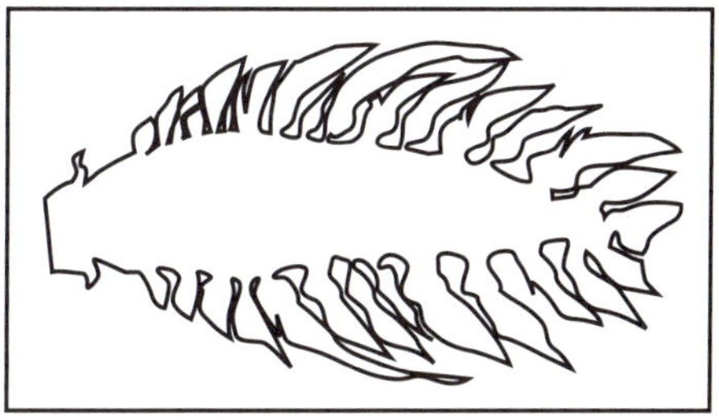

Immature female pine cone

Analysis

1. Compare and **Contrast** the structure and contents of male and female pine cones.

____

____

____

____

2. Critical Thinking

Applying Information It takes 15 months for a pine pollen tube to grow through the wall of a pine ovule. How would you describe the rate of pollen-tube growth in pines?

____

____

Name ______________________ Class ______________ Date ____________

Quick Lab

DATASHEET FOR IN-TEXT LAB

Observing the Arrangement of Parts of a Flower

You can see how the parts of flowers are arranged by dissecting flowers.

MATERIALS

- gloves
- monocot flower
- dicot flower
- paper
- tape

Procedure

1. Put on gloves. Examine a monocot flower and a dicot flower. Locate the sepals, petals, stamens, and pistil of each flower.
2. Separate the parts of each flower, and tape them to a piece of paper. Label each set of parts.
3. Count the number of petals, sepals, and stamens in each flower. Record this information below each flower.

Analysis

1. **Compare** and **Contrast** the appearance of the sepals and petals of each flower.

____

____

2. **Critical Thinking**
Forming a Hypothesis For each flower, suggest a function for the petals based on their appearance.

____

____

____

3. **Critical Thinking**
Justifying Conclusions Explain why each flower is from either a monocot or a dicot.

____

____

____

Name ______________________________ Class ______________ Date ____________

Exploration Lab

DATASHEET FOR IN-TEXT LAB

Observing the Effects of Nutrients on Vegetative Reproduction

SKILLS

- Using scientific processes
- Observing
- Graphing and analyzing data

OBJECTIVES

- **Identify** the structures of duckweed.
- **Compare** vegetative reproduction of duckweed in different nutrient solutions.

MATERIALS

- safety goggles
- lab apron
- duckweed culture
- 5 Petri dishes
- stereomicroscope or hand lens
- glass-marking pen
- beakers
- pond water
- Knop's solution
- 0.1% fertilizer solution
- distilled water

CHEMSAFETY

CAUTION: Always wear safety goggles and a lab apron to protect your eyes and clothing.

CAUTION: Do not touch or taste any chemicals. Know the location of the emergency shower and eyewash station and how to use them. If you get a chemical on your skin or clothing, wash it off at the sink while calling to the teacher. Notify the teacher of a spill. Spills should be cleaned up promptly, according to your teacher's directions.

CAUTION: Glassware is fragile. Notify the teacher of broken glass or cuts. Do not clean up broken glass or spills with broken glass unless the teacher tells you to do so.

Before You Begin

Duckweed is a common aquatic plant. Like many flowering plants, duckweed reproduces readily by **vegetative reproduction,** which is a type of **asexual reproduction.** As individual plants grow, they divide into smaller individuals. Several individuals may remain joined together, forming a mat. All plants require certain **mineral nutrients,** such as nitrogen, phosphorus, and potassium, for the growth of vegetative parts. In this lab, you will investigate the effect of nutrients on the vegetative reproduction of duckweed.

1. Write a definition for each boldface term in the paragraph above. Use a separate sheet of paper.

2. Based on the objectives for this lab, write a question you would like to explore about vegetative reproduction in duckweed.

____

____

Procedure

PART A: MAKE OBSERVATIONS

1. Place a duckweed plant in a Petri dish. Then place a few drops of water on the plant.

2. Observe the duckweed plant with a stereomicroscope or a hand lens. Sketch what you see. Label the structures that you recognize.

PART B: DESIGN AN EXPERIMENT

3. Work with members of your lab group to explore one of the questions written for step 2 of **Before You Begin.** To explore the question, design an experiment that uses the materials listed for this lab.

You Choose

As you design your experiment, decide the following:

a. what question you will explore

b. what your hypothesis will be

c. what solutions to test

d. how much of each solution to use

e. how many individuals to use for each test

f. what your control will be

g. how you will judge which solution is the best

h. what data to record in your data table

4. Write a procedure for your experiment. Make a list of all the safety precautions you will take. Have your teacher approve your procedure and safety precautions before you begin the experiment.

PART C: CONDUCT YOUR EXPERIMENT

5. Put on safety goggles and a lab apron.

6. Set up your experiment. **CAUTION: Nutrient solutions are mild eye irritants. Avoid contact with your skin and eyes.** Complete step 8.

7. Conduct your experiment and collect data for two weeks.

PART D: CLEANUP AND DISPOSAL

8. Dispose of solutions, broken glass, and duckweed in the designated waste containers. Do not pour chemicals down the drain or put lab materials in the trash unless your teacher tells you to do so.

9. Clean up your work area and all lab equipment. Return lab equipment to its proper place. Wash your hands thoroughly before you leave the lab and after you finish all work.

Analyze and Conclude

1. Summarizing Results Compare the appearance of plants growing in each nutrient solution with that of the plants in distilled water. Explain your observations.

____

____

____

____

2. Analyzing Data In which Petri dish did the greatest amount of growth (increase in numbers) take place?

____

3. Analyzing Results In which Petri dish did the least amount of growth take place?

____

4. Evaluating Hypotheses Did the results you observed agree with your hypothesis? If not, how are they different?

____

____

5. **Recognizing Patterns** As the number of new duckweed plants in a particular group increased, what happened to the group of plants?

____

____

6. **Graphing Data** Make a graph of your data. Label the *y*-axis "Number of plants," and the *x*-axis "Days." Use a different color to represent each solution you tested.

____

____

7. **Drawing Conclusions** What factors regulate the rate of vegetative reproduction in duckweed?

____

____

____

8. **Evaluating Methods** Why are the new duckweed plants produced by vegetative reproduction genetically the same as the parent plant?

____

____

____

9. **Further Inquiry** Write a new question about vegetative reproduction in duckweed that could be explored with another investigation.

____

____

Name ______________________ Class ______________ Date ____________

Exploration Lab

OBSERVATION

Fruits and Seeds

The ovary is one of the female reproductive organs of a flower. Often, the ovary consists of two or more compartments called *carpels*. The *ovules*, which mature into *seeds* after fertilization, are found in the carpels. A *fruit* is the ripened ovary of a flower. Because plants need space, sunlight, and nutrients in order to grow, the seeds of most plants must be dispersed or transported away from the parent plant to obtain the proper environment for growth. Many different types of fruit have evolved, with a wide variety of adaptations for the dispersal of their seeds.

In this lab, you will observe and draw a variety of fruits and seeds. You will relate the form of some fruits and seeds to methods of dispersal.

OBJECTIVES

Observe fruits and seeds.

Infer the function of a fleshy fruit.

Compare the ways in which different seeds are dispersed.

MATERIALS

- burrs (sliced open to expose the seeds)
- coconuts (one whole, one split open)
- dissecting microscope
- maple fruits, milkweed pods, or dandelion fruits
- peach or plum (sliced open to expose the pit)
- tomato, apple, cucumber, pumpkin, or green pepper (sliced open across the fruit to expose the seeds)

Procedure

1. Observe a peach or plum. Draw the fruit in the space provided below. Label a seed and the ovary. **CAUTION: You will be working with a variety of fruits and seeds. Do not eat any plant part used in this lab.**

- How many seeds are in the fruit? Describe the location of the seed(s) within the fruit.

  ____

2. Observe a tomato, apple, cucumber, pumpkin, or green pepper. Draw the fruit in the space provided below. Label a seed and the ovary.

- Estimate the number of seeds in the fruit. Describe the location of the seeds within the fruit.

  ____

  ____

- How many carpels (compartments containing seeds) can be seen in the cross-section of your fruit?

  ____

3. Observe maple fruits, milkweed pods, or dandelion fruits. Draw what you see in the space provided below. Label a seed and the ovary wall (if obvious).

- Describe the form and texture of the fruits and seeds. Use a dissecting microscope to help you make your observations.

____

____

____

4. Observe some burrs. Draw an open burr in the space provided below. Label a seed and the ovary.

- Describe the outer covering of the burr.

____

5. Observe a whole coconut. Note the three softer spots at the end of the coconut. Then examine an open coconut.

- Describe the inside of the coconut.

____

- What function do you think the three soft spots might serve?

____

6. Dispose of your materials according to the directions from your teacher.

7. Clean up your work area and wash your hands before leaving the lab.

Analysis

1. Analyzing Data What function do you think the fleshy part of the plum or peach serves?

____

____

2. Identifying Relationships What advantage is provided by the hard covering surrounding a seed?

____

____

3. Analyzing Data How do you think seeds of plants such as maples, dandelions, and milkweeds are dispersed? What features of these seeds led you to your conclusion?

____

____

____

4. Analyzing Data How is the coconut seed that you observed adapted to sea water dispersion?

____

Conclusions

1. Making Predictions Do you think a cucumber or a coconut will have a better chance of producing offspring? Defend your answer.

____

____

____

2. Interpreting Information Which of the seeds that you observed are dispersed by animals? What attributes characterize animal-dispersed seeds?

____

____

Extensions

1. Research and Communications Research the differences between a monocot seed, such as corn, and a dicot seed, such as a lima bean. Dissect, sketch, and label the following parts: cotyledon, embryo, endosperm, seed coat, and hilum. Analyze how the differences between the two types of seeds are beneficial to each type of plant.

2. Research and Communications Discover more about plants by reading about palynology—the study of fossil pollen. Investigate how different species of plants can be identified by their pollen and how pollen grains can remain unchanged for thousands of years.

Name ______________________ Class ______________ Date ____________

Skills Practice Lab

OBSERVATION

Dissecting Flowers

Every flower consists of a set of adaptations that help to ensure successful reproduction. For example, flowers often have bright colors, attractive shapes, and pleasing aromas. These traits help them attract insects and other animals that will carry pollen grains from flower to flower. Pollination also occurs by means other than animals carrying the pollen. For some flowering plants, the wind plays an important role in transferring pollen from plant to plant.

The seed-bearing plants that produce flowers are *angiosperms.* The flower produces the seeds, each of which contains a new plant embryo. The parts of the flower are usually found in four whorls, or rings. *Petals* are one of the four whorls. They attract pollinators. *Sepals* lie outside the petals. They protect the bud.

The reproductive organs, the stamens and pistils, make up the third and fourth whorls and lie inside the petals. A *stamen* is a male reproductive part. It consists of an anther that is held up by a *filament.* The *anther* produces pollen grains. A *pistil* is a female reproductive part. Its top is called the *stigma.* It is sticky to ensure that when pollen grains land on it, they stick to it. The middle supporting structure is the *style*, and the large base is the *ovary*, where the eggs are produced.

In this lab, you will examine a typical flower and observe the parts adapted for the production and protection of seeds.

OBJECTIVES

Identify the structures and arrangements of the female and male reproductive organs of the flower.

Compare and **contrast** sepals and petals.

MATERIALS

- compound light microscope
- coverslip
- dissecting needle
- dropping pipet
- gloves
- hand lens
- microscope slide
- monocot flower or dicot flower
- safety goggles
- scalpel
- water
- sheet of white paper
- transparent tape

Procedure

1. Put on gloves and safety goggles. Examine a sample flower with a hand lens. As you look at the center of your flower, be careful not to damage any parts. Notice how the parts are arranged in circles and attached at the flower base. **CAUTION: Some fluids from plants are irritating to the skin. Be sure to wear gloves when handling plants. Keep your hands away from your face.**

2. Use **Figure 1** to help locate the sepals. The ring of sepals is called the *calyx*. In **Table 1**, record the number of sepals in your flower, describe their appearance, and note the presence or absence of an odor.

FIGURE 1 PARTS OF A FLOWER

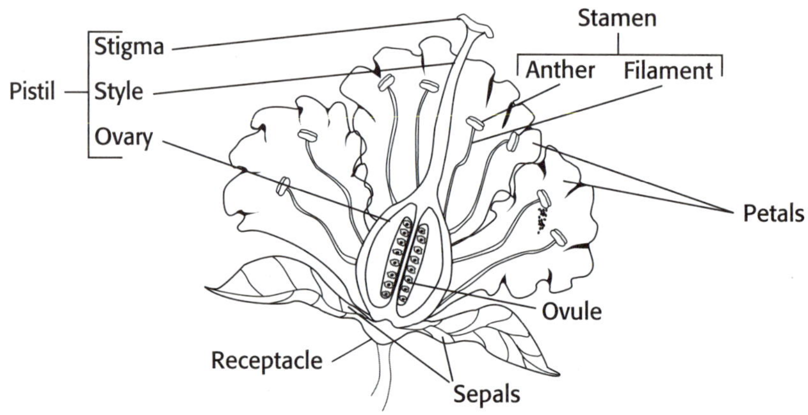

TABLE 1 DESCRIPTION OF FLOWER PARTS

Flower Parts	Description	Odor	Number of Parts
Sepals			
Petals			
Stamens			
Stigma			
Style			
Ovary			
Ovules			

3. Remove the sepals carefully, one by one, by using your first finger and thumb to break off each sepal at the receptacle. Hold a sepal up to the light to observe the veins in it. Place your sepals in a row along the top of a sheet of paper, and attach them using tape. Label the row *Calyx*.

4. The next ring of parts, the petals, is referred to as the *corolla*. In some flowers, the petals are fused at their base. Record the number of petals, their appearance, and the presence or absence of an odor in **Table 1.**

5. Remove the petals carefully, one by one, by using your first finger and thumb. Hold one of the petals up to the light to observe its veins. On your paper just below the sepals, tape your petals in a row. Label this row *Corolla.*
6. Locate the stamens inside the circle of petals. Count and record the number of stamens, and describe their appearance in **Table 1**.
   - In multiples of what number do your stamens occur?

   ____

   ____

7. Gently break off the stamens at their bases. Be careful not to damage the pistil or pistils. On your sheet of paper just below the petals, tape the stamens. Label one stamen, anther, and filament.
8. Make a wet mount of pollen grains taken from one of the anthers. The grains are located along the length of the two pollen sacs of the anther if the grains are mature. If they are not mature, borrow some grains from another lab group. Place a drop of water onto a microscope slide. Use a dissecting needle to scrape off some pollen grains into the water. Place a coverslip over the slide.
9. Examine the slide under a microscope on low power. Then observe it under high power.
10. Observe the pollen grain slide from a lab group with a different type of flower from yours.
    - How do the two types of pollen grains compare?

    ____

    ____

11. Locate the pistil or pistils of the flower. *Note: Your flower will probably have only one pistil, but some flowers have more than one.*
12. Locate and examine the stigma, style, and ovary of one pistil. *Note: Tulips do not have styles.* The ovaries of some flowers (tulips and lilies) are located above the point where the sepals and petals are attached. The ovaries of other flowers (gladioli and irises) are located below the point where the sepals and petals are attached. In **Table 1**, record the appearance of the stigma, style, and ovary.
13. Use a scalpel to cut the ovary open lengthwise up to the style. **CAUTION: Use extreme care with all sharp instruments.** Locate the ovules inside and examine them. In **Table 1**, record the ovules' appearance. Tape the pistil to your sheet of paper below the stamens.
14. Dispose of your materials according to the directions from your teacher.
15. Clean up your work area and wash your hands before leaving the lab.

Analysis

1. **Examining Data** What is the difference between the sepals and petals?

____

____

2. **Explaining Events** How does a pollen grain get to the stigma of a pistil?

____

____

3. **Identifying Relationships** How might the odor of the petals help in pollination?

____

____

4. **Recognizing Patterns** Which parts of the flower develop into the seeds? Which parts develop into the fruits?

____

____

Conclusions

1. **Drawing Conclusions** How does the structure of the stigma aid in pollination?

____

____

2. **Evaluating Data** What type of angiosperm do you have in this lab, a monocot or a dicot? Explain.

____

____

____

Extension

Designing Experiments If your flower was a monocot, repeat this lab using a dicot. If yours was a dicot, use a monocot. Then compare the two kinds of flowers in number of sepals, petals, and stamens. Determine how this information is useful in plant identification.

Name ______________________ Class ______________ Date ______________

Quick Lab

DATASHEET FOR IN-TEXT LAB

Observing a Fern Gametophyte

You can observe the archegonia and antheridia of a fern gametophyte with a microscope.

MATERIALS

- prepared slide of a fern gametophyte with archegonia and antheridia
- compound microscope

Procedure

1. Examine a slide of a fern gametophyte under low power of a microscope. Move the slide until you can see a cluster of archegonia. Now, switch to high power, and focus on one archegonium. Draw and label what you see. Use a separate sheet of paper. **Answers appear in the TE for this lab.**
2. Switch back to low power, and move the slide until you can see several egg-shaped structures. These are antheridia. Now, switch to high power, and focus on one antheridium. Draw and label what you see.

Analysis

1. **Describe** the appearance of an archegonium and an antheridium.

   Answers will vary but should be consistent with the diagrams in Figure 4 of the textbook chapter.

2. **Critical Thinking**
   Drawing Conclusions In which structure, an archegonium or antheridium, does the growth of a new sporophyte begin? Explain.

   A new sporophyte begins its growth in an archegonium. This process occurs because the egg remains in the archegonium, and a sperm cell swims from an antheridium to the archegonium where it fertilizes the egg.

Name ______________________ Class ______________ Date ______________

Quick Lab

DATASHEET FOR IN-TEXT LAB

Observing the Gametophytes of Pines

You can observe the gametophytes of a pine with a microscope.

MATERIALS

- prepared slides of the following: male pine cone, female pine cone, pine ovule
- hand lens
- compound microscope

Procedure

1. Examine prepared slides of male and female pine cones first with a hand lens and then under the low power of a microscope.
2. Make a sketch of each type of pine cone, and label the structures that you recognize.
3. Examine a prepared slide of a pine ovule under the low power of a compound microscope. Compare what you see with the drawing.
4. Draw a pine ovule, and label the following structures: scale, ovule, egg, pollen tube (if visible).

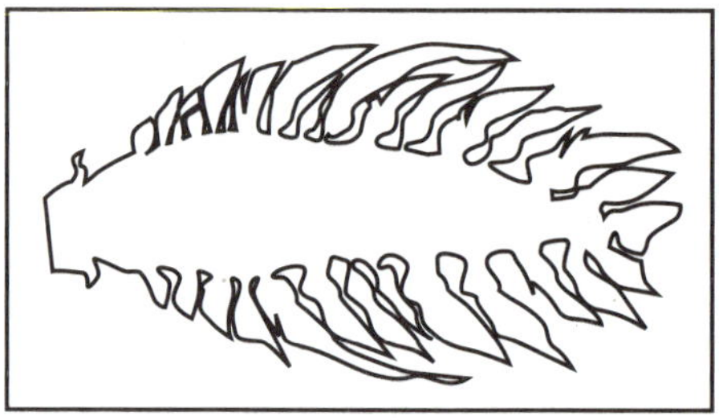

Immature female pine cone

Analysis

1. **Compare** and **Contrast** the structure and contents of male and female pine cones.

   Both have whorls of scales, but male cones are smaller than female cones and contain pollen, while female cones contain eggs and eventually seeds.

2. **Critical Thinking**
   Applying Information It takes 15 months for a pine pollen tube to grow through the wall of a pine ovule. How would you describe the rate of pollen-tube growth in pines?

   The growth rate of pollen tubes in pines is extremely slow.

Name ______________________ Class ______________ Date ______________

Quick Lab

DATASHEET FOR IN-TEXT LAB

Observing the Arrangement of Parts of a Flower

You can see how the parts of flowers are arranged by dissecting flowers.

MATERIALS

- gloves
- monocot flower
- dicot flower
- paper
- tape

Procedure

1. Put on gloves. Examine a monocot flower and a dicot flower. Locate the sepals, petals, stamens, and pistil of each flower.
2. Separate the parts of each flower, and tape them to a piece of paper. Label each set of parts.
3. Count the number of petals, sepals, and stamens in each flower. Record this information below each flower.

Analysis

1. **Compare** and **Contrast** the appearance of the sepals and petals of each flower.

   Answer may vary. Students should comment on color, size, and numbers of sepals and petals.

2. **Critical Thinking**
   Forming a Hypothesis For each flower, suggest a function for the petals based on their appearance.

   Answers will vary. Students should suggest that large, brightly colored petals may attract animal pollinators (such as insects) while flowers with small, inconspicuous petals are probably wind-pollinated.

3. **Critical Thinking**
   Justifying Conclusions Explain why each flower is from either a monocot or a dicot.

   Answers may vary. Students should note the following: monocots have flower parts in multiples of three; dicots have flower parts in multiples of two, four, or five.

Name ______________________ Class ______________ Date ____________

Exploration Lab

DATASHEET FOR IN-TEXT LAB

Observing the Effects of Nutrients on Vegetative Reproduction

SKILLS

- Using scientific processes
- Observing
- Graphing and analyzing data

OBJECTIVES

- **Identify** the structures of duckweed.
- **Compare** vegetative reproduction of duckweed in different nutrient solutions.

MATERIALS

- safety goggles
- lab apron
- duckweed culture
- 5 Petri dishes
- stereomicroscope or hand lens
- glass-marking pen
- beakers
- pond water
- Knop's solution
- 0.1% fertilizer solution
- distilled water

CHEMSAFETY

CAUTION: Always wear safety goggles and a lab apron to protect your eyes and clothing.

CAUTION: Do not touch or taste any chemicals. Know the location of the emergency shower and eyewash station and how to use them. If you get a chemical on your skin or clothing, wash it off at the sink while calling to the teacher. Notify the teacher of a spill. Spills should be cleaned up promptly, according to your teacher's directions.

CAUTION: Glassware is fragile. Notify the teacher of broken glass or cuts. Do not clean up broken glass or spills with broken glass unless the teacher tells you to do so.

Before You Begin

Duckweed is a common aquatic plant. Like many flowering plants, duckweed reproduces readily by **vegetative reproduction,** which is a type of **asexual reproduction.** As individual plants grow, they divide into smaller individuals. Several individuals may remain joined together, forming a mat. All plants require certain **mineral nutrients,** such as nitrogen, phosphorus, and potassium, for the growth of vegetative parts. In this lab, you will investigate the effect of nutrients on the vegetative reproduction of duckweed.

1. Write a definition for each boldface term in the paragraph above. Use a separate sheet of paper. **Answers appear in the TE for this lab.**
2. Based on the objectives for this lab, write a question you would like to explore about vegetative reproduction in duckweed.

   For example: Which nutrients are needed for vegetative reproduction in duckweed?

Procedure

PART A: MAKE OBSERVATIONS

1. Place a duckweed plant in a Petri dish. Then place a few drops of water on the plant.
2. Observe the duckweed plant with a stereomicroscope or a hand lens. Sketch what you see. Label the structures that you recognize.

PART B: DESIGN AN EXPERIMENT

3. Work with members of your lab group to explore one of the questions written for step 2 of **Before You Begin.** To explore the question, design an experiment that uses the materials listed for this lab.

You Choose

As you design your experiment, decide the following:

- **a.** what question you will explore
- **b.** what your hypothesis will be
- **c.** what solutions to test
- **d.** how much of each solution to use
- **e.** how many individuals to use for each test
- **f.** what your control will be
- **g.** how you will judge which solution is the best
- **h.** what data to record in your data table

Name ________________________ Class ______________ Date ______________

Observing the Effects of Nutrients on Vegetative Reproduction *continued*

4. Write a procedure for your experiment. Make a list of all the safety precautions you will take. Have your teacher approve your procedure and safety precautions before you begin the experiment.

PART C: CONDUCT YOUR EXPERIMENT

5. Put on safety goggles and a lab apron.

6. Set up your experiment. **CAUTION: Nutrient solutions are mild eye irritants. Avoid contact with your skin and eyes.** Complete step 8.

7. Conduct your experiment and collect data for two weeks.

PART D: CLEANUP AND DISPOSAL

8. Dispose of solutions, broken glass, and duckweed in the designated waste containers. Do not pour chemicals down the drain or put lab materials in the trash unless your teacher tells you to do so.

9. Clean up your work area and all lab equipment. Return lab equipment to its proper place. Wash your hands thoroughly before you leave the lab and after you finish all work.

Analyze and Conclude

1. Summarizing Results Compare the appearance of plants growing in each nutrient solution with that of the plants in distilled water. Explain your observations.

Plants grown in Knop's solution should be the largest and darkest green. Plants grown in pond water should also look healthy and green. Plants grown in dilute fertilizer solution or distilled water should be smaller and lighter green.

2. Analyzing Data In which Petri dish did the greatest amount of growth (increase in numbers) take place?

The greatest amount of growth should occur in the dish with Knop's solution.

3. Analyzing Results In which Petri dish did the least amount of growth take place?

The least amount of growth should occur in distilled water.

4. Evaluating Hypotheses Did the results you observed agree with your hypothesis? If not, how are they different?

Answers will vary. Students should state how their results differ from their hypotheses.

Name ______________________ Class ______________ Date ____________

Observing the Effects of Nutrients on Vegetative Reproduction *continued*

5. **Recognizing Patterns** As the number of new duckweed plants in a particular group increased, what happened to the group of plants?

The groups divide as they get larger.

6. **Graphing Data** Make a graph of your data. Label the *y*-axis "Number of plants," and the *x*-axis "Days." Use a different color to represent each solution you tested.

Answer will vary.

7. **Drawing Conclusions** What factors regulate the rate of vegetative reproduction in duckweed?

Students should recognize that more mineral nutrients will produce more growth.

8. **Evaluating Methods** Why are the new duckweed plants produced by vegetative reproduction genetically the same as the parent plant?

Vegetative reproduction involves mitotic cell division, and mitosis results in the production of genetically identical cells.

9. **Further Inquiry** Write a new question about vegetative reproduction in duckweed that could be explored with another investigation.

Answers will vary.

Exploration Lab

OBSERVATION

Fruits and Seeds

Teacher Notes

TIME REQUIRED One 45-minute period

SKILLS ACQUIRED

Collecting data
Identifying patterns
Inferring
Interpreting
Organizing and analyzing data

RATING

Easy ← 1 2 3 4 → Hard

Teacher Prep–2
Student Setup–1
Concept Level–2
Cleanup–1

THE SCIENTIFIC METHOD

Make Observations Students observe a variety of fruits and seeds.

Analyze the Results Analysis questions 1, 3, and 4 require students to analyze their results.

Draw Conclusions Conclusions questions 1 and 2 require students to draw conclusions from their data.

MATERIALS

Materials for this lab can be purchased from WARD'S. See the *Master Materials List* for ordering instructions.

For step 1, any single-seeded fleshy fruit (olive, avocado, or cherry) can be used. For step 2, a green bean or squash will also work. For step 3, any wind-carried fruit or seed, such as ailanthus or thistle, will work. Though burrs work best for step 4, fruits and seeds from beggar's ticks, Queen Anne's lace, or tick trefoil can be substituted.

SAFETY CAUTIONS

- Discuss all safety symbols and caution statements with students.
- Remind students not to eat the fruits and seeds used during the lab.

TIPS AND TRICKS

This lab works best in groups of two students.

Pre-slice all fruits to be used in this lab. You may want to cut fruits used in step 2 along different axes to help students get a three-dimensional idea of the shape of the ovary. While wearing safety goggles, use a hammer and chisel to split open one coconut. Crack open some of the burrs so that the seeds can be seen.

Name ______________________ Class ______________ Date ____________

Exploration Lab

OBSERVATION

Fruits and Seeds

The ovary is one of the female reproductive organs of a flower. Often, the ovary consists of two or more compartments called *carpels*. The *ovules*, which mature into *seeds* after fertilization, are found in the carpels. A *fruit* is the ripened ovary of a flower. Because plants need space, sunlight, and nutrients in order to grow, the seeds of most plants must be dispersed or transported away from the parent plant to obtain the proper environment for growth. Many different types of fruit have evolved, with a wide variety of adaptations for the dispersal of their seeds.

In this lab, you will observe and draw a variety of fruits and seeds. You will relate the form of some fruits and seeds to methods of dispersal.

OBJECTIVES

Observe fruits and seeds.

Infer the function of a fleshy fruit.

Compare the ways in which different seeds are dispersed.

MATERIALS

- burrs (sliced open to expose the seeds)
- coconuts (one whole, one split open)
- dissecting microscope
- maple fruits, milkweed pods, or dandelion fruits
- peach or plum (sliced open to expose the pit)
- tomato, apple, cucumber, pumpkin, or green pepper (sliced open across the fruit to expose the seeds)

Procedure

1. Observe a peach or plum. Draw the fruit in the space provided below. Label a seed and the ovary. **CAUTION: You will be working with a variety of fruits and seeds. Do not eat any plant part used in this lab.**

Drawings will vary depending on the fruit used.

- How many seeds are in the fruit? Describe the location of the seed(s) within the fruit.

  The fruit has one seed. It is inside the pit at the center of the fruit.

2. Observe a tomato, apple, cucumber, pumpkin, or green pepper. Draw the fruit in the space provided below. Label a seed and the ovary.

Drawings will vary depending on the fruit used.

- Estimate the number of seeds in the fruit. Describe the location of the seeds within the fruit.

  Answers will vary depending on the fruit, but there will be many seeds in the fruits. The seeds will be arranged radially in compartments.

- How many carpels (compartments containing seeds) can be seen in the cross-section of your fruit?

  Answers will vary but will most likely be 3 to 5 carpels.

3. Observe maple fruits, milkweed pods, or dandelion fruits. Draw what you see in the space provided below. Label a seed and the ovary wall (if obvious).

Drawings will vary depending on the seed used.

Name ______________________ Class ______________ Date ____________

- Describe the form and texture of the fruits and seeds. Use a dissecting microscope to help you make your observations.

  Maple fruits have a stiff wing attached to the seed cover. Dandelion fruits and milkweed seeds have light, feathery outgrowths that act as parachutes.

4. Observe some burrs. Draw an open burr in the space provided below. Label a seed and the ovary.

Drawings should indicate tiny seeds inside the burr and the inner wall of the burr as the ovary.

- Describe the outer covering of the burr.

  The burr is covered with sharp, hooklike structures.

5. Observe a whole coconut. Note the three softer spots at the end of the coconut. Then examine an open coconut.

- Describe the inside of the coconut.

  It is hollow with a fleshy layer covering the inside of the shell.

- What function do you think the three soft spots might serve?

  When the seed germinates, the new plant emerges from one of the spots.

6. Dispose of your materials according to the directions from your teacher.

7. Clean up your work area and wash your hands before leaving the lab.

Analysis

1. Analyzing Data What function do you think the fleshy part of the plum or peach serves?

The fleshy part protects the seed and attracts animals that eat the fruit and disperse the seeds.

Name ______________________ Class ______________ Date ____________

2. **Identifying Relationships** What advantage is provided by the hard covering surrounding a seed?

   The seed is protected from dehydration and other damage until the conditions for germination are favorable.

3. **Analyzing Data** How do you think seeds of plants such as maples, dandelions, and milkweeds are dispersed? What features of these seeds led you to your conclusion?

   The seeds are dispersed by the wind. The wing shape of the maple fruits and the parachutelike plumes of the lightweight dandelion and milkweed seeds are telltale features.

4. **Analyzing Data** How is the coconut seed that you observed adapted to sea water dispersion?

   It is hollow and light and thus can float.

Conclusions

1. **Making Predictions** Do you think a cucumber or a coconut will have a better chance of producing offspring? Defend your answer.

   Students may say the cucumber because it contains many more seeds than a coconut. Producing a large number of seeds increases the chances that a viable seed will land in a suitable place to grow and thus help the species to survive.

2. **Interpreting Information** Which of the seeds that you observed are dispersed by animals? What attributes characterize animal-dispersed seeds?

   Answers may include sweet fruit to make the animal want to eat it and stickiness or hooklike structures to make the seeds or fruit cling to an animal.

Extensions

1. **Research and Communications** Research the differences between a monocot seed, such as corn, and a dicot seed, such as a lima bean. Dissect, sketch, and label the following parts: cotyledon, embryo, endosperm, seed coat, and hilum. Analyze how the differences between the two types of seeds are beneficial to each type of plant.
2. **Research and Communications** Discover more about plants by reading about palynology—the study of fossil pollen. Investigate how different species of plants can be identified by their pollen and how pollen grains can remain unchanged for thousands of years.

Skills Practice Lab

OBSERVATION

Dissecting Flowers

Teacher Notes

TIME REQUIRED One 45-minute period

SKILLS ACQUIRED

Collecting data
Experimenting
Identifying patterns
Inferring
Interpreting
Measuring
Organizing and analyzing data

RATING Easy ← 1 2 3 4 → Hard

Teacher Prep–3
Student Setup–2
Concept Level–2
Cleanup–2

THE SCIENTIFIC METHOD

Make Observations Students make observations about the structures of a flower.

Analyze the Results Analysis questions 1-4 require students to analyze their results.

Draw Conclusions Conclusions questions 1 and 2 ask students to draw conclusions from their data.

MATERIALS

Many materials for this lab can be purchased from WARD'S. See the *Master Materials List* for ordering instructions.

Monocots may include gladiolus, tulip, or lily; dicots may be azalea, camelia, evening primrose, morning glory, or snapdragon.

Have models of monocots and dicots available for use by those students who have negative topical or respiratory reactions to flowers.

SAFETY CAUTIONS

- Discuss all safety symbols and caution statements with students.
- Caution students not to hold the pistil or ovaries in their hand when slicing them open. Provide a suitable cutting surface on which to perform this task.
- Check with students ahead of this lab session to make sure none of them have allergies to any of the flowers that might be used.

DISPOSAL

Discarded flower parts may be placed in the laboratory trash can.

TIPS AND TRICKS

This lab works best in groups of two students.

Before the lab, call florists, funeral homes, or grocery stores that carry fresh flowers to arrange for donations of flowers. You can also have students bring in flowers for this lab. If the school has a horticulture department, they might be able to donate flowers.

It is possible that students will be trying to dissect a flower with an inferior ovary, e.g., gladiolus or evening primrose. Inferior ovaries are located below, or embedded in, the receptacle.

Name ______________________ Class ______________ Date ____________

Skills Practice Lab

OBSERVATION

Dissecting Flowers

Every flower consists of a set of adaptations that help to ensure successful reproduction. For example, flowers often have bright colors, attractive shapes, and pleasing aromas. These traits help them attract insects and other animals that will carry pollen grains from flower to flower. Pollination also occurs by means other than animals carrying the pollen. For some flowering plants, the wind plays an important role in transferring pollen from plant to plant.

The seed-bearing plants that produce flowers are *angiosperms.* The flower produces the seeds, each of which contains a new plant embryo. The parts of the flower are usually found in four whorls, or rings. *Petals* are one of the four whorls. They attract pollinators. *Sepals* lie outside the petals. They protect the bud.

The reproductive organs, the stamens and pistils, make up the third and fourth whorls and lie inside the petals. A *stamen* is a male reproductive part. It consists of an anther that is held up by a *filament.* The *anther* produces pollen grains. A *pistil* is a female reproductive part. Its top is called the *stigma.* It is sticky to ensure that when pollen grains land on it, they stick to it. The middle supporting structure is the *style,* and the large base is the *ovary,* where the eggs are produced.

In this lab, you will examine a typical flower and observe the parts adapted for the production and protection of seeds.

OBJECTIVES

Identify the structures and arrangements of the female and male reproductive organs of the flower.

Compare and **contrast** sepals and petals.

MATERIALS

- compound light microscope
- coverslip
- dissecting needle
- dropping pipet
- gloves
- hand lens
- microscope slide
- monocot flower or dicot flower
- safety goggles
- scalpel
- water
- sheet of white paper
- transparent tape

Name ______________________ Class ______________ Date ______________

Procedure

1. Put on gloves and safety goggles. Examine a sample flower with a hand lens. As you look at the center of your flower, be careful not to damage any parts. Notice how the parts are arranged in circles and attached at the flower base. **CAUTION: Some fluids from plants are irritating to the skin. Be sure to wear gloves when handling plants. Keep your hands away from your face.**

2. Use **Figure 1** to help locate the sepals. The ring of sepals is called the *calyx*. In **Table 1**, record the number of sepals in your flower, describe their appearance, and note the presence or absence of an odor.

FIGURE 1 PARTS OF A FLOWER

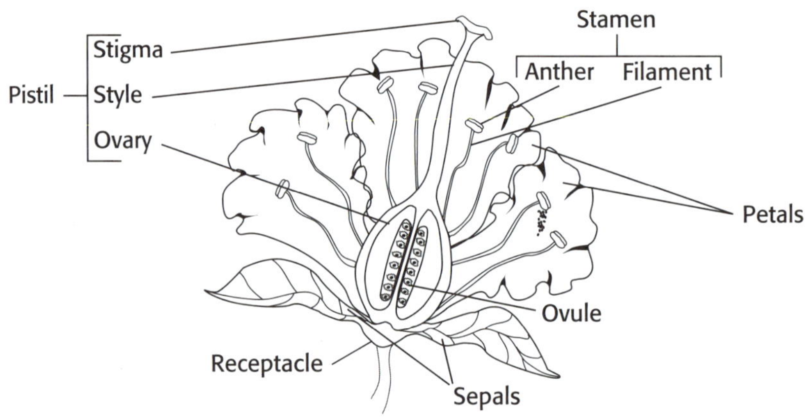

TABLE 1 DESCRIPTION OF FLOWER PARTS

Flower Parts	Description	Odor	Number of Parts
Sepals			
Petals			
Stamens	**Answers in table will vary.**		
Stigma			
Style			
Ovary			
Ovules			

3. Remove the sepals carefully, one by one, by using your first finger and thumb to break off each sepal at the receptacle. Hold a sepal up to the light to observe the veins in it. Place your sepals in a row along the top of a sheet of paper, and attach them using tape. Label the row *Calyx*.

4. The next ring of parts, the petals, is referred to as the *corolla*. In some flowers, the petals are fused at their base. Record the number of petals, their appearance, and the presence or absence of an odor in **Table 1.**

Name ______________________ Class ______________ Date ____________

Dissecting Flowers *continued*

5. Remove the petals carefully, one by one, by using your first finger and thumb. Hold one of the petals up to the light to observe its veins. On your paper just below the sepals, tape your petals in a row. Label this row *Corolla.*
6. Locate the stamens inside the circle of petals. Count and record the number of stamens, and describe their appearance in **Table 1**.
   - In multiples of what number do your stamens occur?

     Answers will vary. Monocots have flower parts in multiples of three, and dicots have flower parts in multiples of two, four, or five.
7. Gently break off the stamens at their bases. Be careful not to damage the pistil or pistils. On your sheet of paper just below the petals, tape the stamens. Label one stamen, anther, and filament.
8. Make a wet mount of pollen grains taken from one of the anthers. The grains are located along the length of the two pollen sacs of the anther if the grains are mature. If they are not mature, borrow some grains from another lab group. Place a drop of water onto a microscope slide. Use a dissecting needle to scrape off some pollen grains into the water. Place a coverslip over the slide.
9. Examine the slide under a microscope on low power. Then observe it under high power.
10. Observe the pollen grain slide from a lab group with a different type of flower from yours.
    - How do the two types of pollen grains compare?

      Answers will vary because different species of plants may produce different shapes of pollen.
11. Locate the pistil or pistils of the flower. *Note: Your flower will probably have only one pistil, but some flowers have more than one.*
12. Locate and examine the stigma, style, and ovary of one pistil. *Note: Tulips do not have styles.* The ovaries of some flowers (tulips and lilies) are located above the point where the sepals and petals are attached. The ovaries of other flowers (gladioli and irises) are located below the point where the sepals and petals are attached. In **Table 1**, record the appearance of the stigma, style, and ovary.
13. Use a scalpel to cut the ovary open lengthwise up to the style. **CAUTION: Use extreme care with all sharp instruments.** Locate the ovules inside and examine them. In **Table 1**, record the ovules' appearance. Tape the pistil to your sheet of paper below the stamens.
14. Dispose of your materials according to the directions from your teacher.
15. Clean up your work area and wash your hands before leaving the lab.

Analysis

1. **Examining Data** What is the difference between the sepals and petals?

   Sepals are often green, although they may be other colors, and are very thick at the base. Petals may be brightly colored and are inside the sepals.

2. **Explaining Events** How does a pollen grain get to the stigma of a pistil?

   A pollen grain is generally moved to the stigma by animals or wind, or it simply falls on the stigma.

3. **Identifying Relationships** How might the odor of the petals help in pollination?

   Some animals that act as pollinators are attracted by the odors.

4. **Recognizing Patterns** Which parts of the flower develop into the seeds? Which parts develop into the fruits?

   The ovules become seeds, and the ovary (and sometimes the receptacle) becomes the fruit.

Conclusions

1. **Drawing Conclusions** How does the structure of the stigma aid in pollination?

   The stigma projects up from the ovary so as to be exposed, and it has a sticky surface that pollen grains will easily cling to.

2. **Evaluating Data** What type of angiosperm do you have in this lab, a monocot or a dicot? Explain.

   Answers will vary. If the student counted the stamens, sepals, or petals, and recorded a multiple of three, it is probably a monocot. If the student recorded a multiple of two, four, or five, it is probably a dicot.

Extension

Designing Experiments If your flower was a monocot, repeat this lab using a dicot. If yours was a dicot, use a monocot. Then compare the two kinds of flowers in number of sepals, petals, and stamens. Determine how this information is useful in plant identification.

Answer Key

Directed Reading

SECTION: SEXUAL REPRODUCTION IN SEEDLESS PLANTS

1. 1
2. 4
3. 5
4. 3
5. 2
6. 6
7. 1
8. 2
9. 5
10. 4
11. 3

SECTION: SEXUAL REPRODUCTION IN SEED PLANTS

1. pollen grain
2. ovule
3. pollination
4. pollen tube
5. A seed coat is the protective outer covering of a seed.
6. The seed coat protects the embryo in a seed from mechanical injury and prevents the embryo from growing during unfavorable weather by keeping out water and oxygen.
7. Cotyledons are leaflike structures that are a part of the plant embryo in a seed.
8. Cotyledons help transfer nutrients to the embryo.
9. 1
10. 2
11. 6
12. 3
13. 5
14. 4
15. c
16. d
17. e
18. b
19. a
20. d
21. b

SECTION: ASEXUAL REPRODUCTION

1. A bulb is a modified stem that has thick, fleshy leaves. A corm is a modified stem that has thin, scaly leaves.
2. A tuber is a swollen and fleshy underground stem. A rhizome is a horizontal underground stem that is not swollen and fleshy.
3. A plantlet, which is a means of vegetative reproduction in kalanchoës, is a tiny new plant. A flower is a sexual reproductive organ in an angiosperm.
4. Plant propagation is the growing of new plants from seed or vegetative parts.
5. In grafting, pieces cut from one plant are attached to another plant. Taking cuttings involves cutting off pieces of a plant and using them to grow whole new individuals.

Active Reading

SECTION: SEXUAL REPRODUCTION IN SEEDLESS PLANTS

1. Through meiosis, a diploid sporophyte produces spores.
2. Through mitosis, fern gametophytes produce gametes.
3. Sperm can fertilize the eggs inside the archegonia.
4. 4
5. 2
6. 5
7. 1
8. 3
9. adult sporophyte
10. spores
11. mature gametophyte
12. mitosis
13. archegonium
14. fertilization
15. zygote
16. sporangium
17. c

SECTION: SEXUAL REPRODUCTION IN SEED PLANTS

1. in four concentric whorls
2. to protect the flower from damage while it is a bud
3. to attract pollinators to the flower
4. An anther is a pollen-producing sac located at the tip of a stamen.
5. a

SECTION: ASEXUAL REPRODUCTION

1. growing new plants from seed or vegetative parts
2. grafting, taking cuttings, and tissue culture
3. a description of each method as well

as examples of plants commonly produced by each type of vegetative plant propagation
4. small stems from one plant are attached to larger stems or roots of another plant
5. African violets, ornamental trees and shrubs, and figs
6. b

Vocabulary Review

1. c
2. b
3. o
4. k
5. g
6. j
7. l
8. m
9. d
10. n
11. h
12. p
13. a
14. i
15. f
16. e
17. vegetative reproduction
18. plant propagation
19. tissue culture

Science Skills

CLASSIFYING/SEQUENCING

1. **Nonvascular plants**
   swimming sperm
   independent gametophytes
   spores
   antheridia and archegonia
2. **Seedless vascular plants**
   swimming sperm
   independent gametophytes
   true roots and leaves
   spores
   antheridia and archegonia
3. **Gymnosperms**
   pollen
   true roots and leaves
   independent sporophytes
   spores
   pollination by wind
   seeds
   gametophytes inside male cones and female cones
4. **Angiosperms**
   pollen
   flowers
   true roots and leaves
   independent sporophytes
   spores
   pollination by wind
   seeds
   pollination by animals
5. 2
6. 5
7. 4
8. 3
9. 6
10. 1

Concept Mapping

1. sexual reproduction
2. asexual reproduction
3. flowers or cones
4. cones or flowers
5. taking cuttings or tissue culture or grafting
6. tissue culture or grafting or taking cuttings
7. grafting or taking cuttings or tissue culture
8. bulbs
9. corms
10. rhizomes
11. runners

Answers for 8–11 are interchangeable.

Critical Thinking

1. d
2. a
3. b
4. c
5. e
6. b
7. a
8. c
9. d
10. b
11. a
12. c
13. b, h
14. c, a
15. d, e
16. f, g
17. c
18. a
19. b
20. d

Test Prep Pretest

1. c
2. b
3. a
4. c
5. c
6. b
7. c
8. a
9. g
10. f
11. b
12. e
13. d
14. a
15. c
16. sporangium, sorus
17. pollen grains, ovule
18. archegonia, antheridia
19. spores, pistils, stamens
20. rhizomes, vegetative

21. triploid, endosperm
22. In both, a haploid gametophyte that produces gametes by mitosis alternates with a diploid sporophyte that produces spores by meiosis. However, in a moss, the gametophyte is larger than the sporophyte. The reverse is true in the fern. Both mosses and ferns require water for fertilization to occur.
23. sporophyte
24. sporangium
25. gametophyte

Quiz

SECTION: SEXUAL REPRODUCTION IN SEEDLESS PLANTS

1. a
2. c
3. e
4. d
5. a
6. b
7. a
8. c
9. a
10. a

SECTION: SEXUAL REPRODUCTION IN SEED PLANTS

1. a
2. d
3. c
4. b
5. e
6. a
7. c
8. a
9. a
10. c

SECTION: ASEXUAL REPRODUCTION

1. d
2. b
3. c
4. e
5. f
6. a
7. b
8. a
9. d
10. c

Chapter Test (General)

1. a, f
2. f
3. b, d, e
4. c, d
5. a
6. d
7. b
8. b
9. c
10. d
11. c
12. d
13. c
14. c
15. a
16. d
17. a
18. c
19. d
20. d

Chapter Test (Advanced)

1. d
2. a
3. c
4. b
5. d
6. c
7. b
8. c
9. b
10. a
11. d
12. b
13. c
14. gametophytes, pistil
15. pollen, pollen tube, ovule
16. cotyledon
17. sepals, petals, stamens, pistil
18. seed, sporophyte
19. imperfect
20. seeds, vegetative parts
21. The diploid sporophyte produces haploid spores by meiosis. The spores fall to the ground and grow into haploid gametophytes, which produce gametes—sperm and eggs—by mitosis. When water covers the gametophytes, sperm can swim to the archegonia and fertilize the eggs inside of them. The zygotes, or fertilized eggs, grow into new sporophytes.
22. At the time of pollination, the scales of a female cone are open, exposing the ovules. When a pollen grain lands near an ovule, a slender pollen tube grows out of the pollen grain and into the ovule. The sperm moves through the pollen tube and enters the ovule to reach the egg. The pollen tube provides a way for the sperm to reach the egg without having to swim through water.
23. The pollen tube links a pollen grain with an ovule. Each grain contains two sperm that travel through the tube to the ovule. One sperm fuses with the egg and forms the zygote. The other sperm fuses with two other haploid nuclei and develops into the endosperm. The zygote and the endosperm develop into a seed, which grows into a new sporophyte.
24. Many plants produce structures that can grow into new plants. Some examples are bulbs, corms, rhizomes, and tubers, which are modified stems. Pieces of roots, stems, and leaves may also grow into new plants.
25. Plantlets are tiny new plants that develop by vegetative reproduction in notches along the leaf margins of the kalanchoë. When a plantlet falls to the ground, it grows into a plant that is identical to the parent plant.

Lesson Plan

Section: Sexual Reproduction in Seedless Plants

Pacing

Regular Schedule:	**with lab(s):** N/A	**without lab(s):** 3 days
Block Schedule:	**with lab(s):** N/A	**without lab(s):** 1 1/2 days

Objectives

1. Summarize the life cycle of a moss.
2. Summarize the life cycle of a fern.
3. Compare and contrast the life cycle of a moss with the life cycle of a fern.

National Science Education Standards Covered

UNIFYING CONCEPTS AND PROCESSES

UCP1: Systems, order, and organization

UCP2: Evidence, models, and explanation

UCP5: Form and function

SCIENCE AS INQUIRY

SI1: Abilities necessary to do scientific inquiry

SI2: Understandings about scientific inquiry

LIFE SCIENCE: THE CELL

LSCell1: Cells have particular structures that underlie their functions.

LSCell4: Cell functions are regulated.

LSCell5: Plant cells contain chloroplasts, the site of photosynthesis.

LSCell6: Cells can differentiate and form complete multicellular organisms.

KEY
SE = Student Edition — **TE** = Teacher Edition
CRF = Chapter Resource File

Block 1

CHAPTER OPENER *(45 minutes)*

_ **Quick Review,** SE. Students answer questions covered in previous sections of the textbook as preparation for the chapter content. **(GENERAL)**

_ **Reading Activity,** SE. Students write down all the key words for each section of the chapter and write a definition next to each word that they have heard of. Then as students read, they write definitions next to the remaining words. **(GENERAL)**

_ **Using the Figure,** TE. Students answer questions about the chapter opener photograph. **(GENERAL)**

_ **Opening Activity**, Plant Display, TE. Display some plants with sexual and asexual reproductive structures. Discuss the form of reproduction each structure represents.

Block 2

FOCUS *(5 minutes)*

_ **Bellringer Transparency.** Use this transparency as students enter the classroom and find their seats. **(GENERAL)**

MOTIVATE *(10 minutes)*

_ **Discussion/Question**, TE. Tell students that the sperm of nonvascular plants must swim to an egg for fertilization to occur. Ask students to describe the environmental conditions that allow nonvascular plants to reproduce sexually.

TEACH *(30 minutes)*

_ **Teaching Transparency, Section Outline.** Use this transparency to give students a framework for the information in this section. **(GENERAL)**

_ **Directed Reading Worksheet, Sexual Reproduction in Seedless Plants, CRF.** Students complete the exercises in this worksheet to help them understand the material as they read the section. **(BASIC)**

_ **Teaching Transparency, Life Cycle of a Moss.** Use this transparency to discuss the life cycle of a moss. Point out that when meiosis occurs, the chromosome number changes from diploid to haploid. Also point out that a spore rows into a gametophyte, which has haploid cells. Note that when fertilization occurs, the chromosome number becomes diploid and a new sporophyte generation begins. **(GENERAL)**

_ **Using the Figure**, Figure 2, TE. Use the suggestions in this TE item to walk students through the fern life cycle, starting with the zygote.

HOMEWORK

_ **Active Reading Worksheet, Sexual Reproduction in Seedless Plants, CRF.** Students read a passage related to the section topic and answer questions. **(GENERAL)**

Block 3

TEACH *(25 minutes)*

_ **Quick Lab,** Observing a Fern Gametophyte, SE. Students study a fern gametophyte under a microscope and describe the appearance of the archegonia and antheridia. **(GENERAL)**

_ **Datasheets for In-Text Labs, Observing a Fern Gametophyte, CRF.**

_ **Teaching Transparency, Life Cycle of a Fern.** Use this transparency to discuss the the life cycle of a fern. Point out that when meiosis occurs, the chromosome number changes from diploid to haploid. Also point out that a spore grows into a gametophyte, which is haploid. Note that when fertilization occurs, the chromosome number becomes diploid and a new sporophyte generation begins. **(GENERAL)**

CLOSE *(20 minutes)*

_ **Reteaching,** TE. Summarize the life cycles of mosses and ferns. Emphasize that all plants alternate between haploid and diploid phases in their life cycles. **(BASIC)**

_ **Alternative Assessment**, TE. Students work in groups to make posters that depict the life cycles of mosses and ferns. **(GENERAL)**

_ **Quiz,** TE. Students answer questions that review the section material. **(GENERAL)**

HOMEWORK

_ **Section Review,** SE. Assign questions 1–5 for review, homework, or quiz. **(GENERAL)**

_ **Quiz, CRF.** This quiz consists of ten multiple choice and matching questions that review the section's main concepts. **(BASIC) Also in Spanish.**

Other Resource Options

_ **Misconception Alert,** Spores, TE. Discuss the meaning of the word *spore* in different biological contexts.

_ **go.hrw.com.** For worksheets, videos, and other teaching aids related to this chapter, visit the HRW Web site and type in the keyword HX4 PRP.

_ **CNN Science in the News, Video Segment 20 Migrant Bees.** This video segment is accompanied by a **Critical Thinking Worksheet**.

_ **CNN Student News.** Find the latest news, lesson plans, and activities related to important scientific events at **cnnstudentnews.com**.

Lesson Plan

Section: Sexual Reproduction in Seed Plants

Pacing

Regular Schedule:	**with lab(s):** 4 days	**without lab(s):** 2 days
Block Schedule:	**with lab(s):** 2 days	**without lab(s):** 1 day

Objectives

1. Distinguish the male and the female gametophytes of seed plants.
2. Describe the function of each part of a seed.
3. Summarize the life cycle of a conifer.
4. Relate the parts of a flower to their functions.
5. Summarize the life cycle of an angiosperm.

National Science Education Standards Covered

UNIFYING CONCEPTS AND PROCESSES

UCP1: Systems, order, and organization

UCP2: Evidence, models, and explanation

UCP5: Form and function

SCIENCE AS INQUIRY

SI1: Abilities necessary to do scientific inquiry

SI2: Understandings about scientific inquiry

LIFE SCIENCE: THE CELL

LSCell1: Cells have particular structures that underlie their functions.

LSCell4: Cell functions are regulated.

LSCell5: Plant cells contain chloroplasts, the site of photosynthesis.

LSCell6: Cells can differentiate and form complete multicellular organisms.

LIFE SCIENCE: INTERDEPENDENCE OF ORGANISMS

LSInter3: Organisms both cooperate and compete in ecosystems.

KEY
SE = Student Edition **TE** = Teacher Edition
CRF = Chapter Resource File

Block 4

FOCUS *(5 minutes)*

_ **Bellringer Transparency.** Use this transparency as students enter the classroom and find their seats. **(GENERAL)**

MOTIVATE *(10 minutes)*

_ **Activity**, Gymnosperm Flash Cards, TE. Students write each key terms from this section on a separate note card along with its definition (on the back) as they read the section.

TEACH *(30 minutes)*

_ **Teaching Transparency, Section Outline.** Use this transparency to give students a framework for the information in this section. **(GENERAL)**

_ **Teaching Transparency, Seed Plant Gametophyte.** Use this transparency to discuss how gametophyets of seed plants are different from those of seedless plants. Review the parts of a flower. **(GENERAL)**

_ **Teaching Transparency, Seed Structure.** Use this transparency to compare the structures of a pine seed, corn grain, and bean seed. **(GENERAL)**

_ **Group Activity**, Dissecting Seeds, TE. Bring a variety of seeds to class and make a longitudinal cut through at least one seed of each type. Have students examine whole and opened seeds using their unaided eyes and a dissecting scope. Ask students to compare the characteristics of the different types of seeds.

_ **Teaching Transparency, Life Cycle of a Conifer.** Use this transparency to discuss the life cycle of conifers. Point out that when meiosis occurs, the chromosome number changes from diploid to haploid. **(GENERAL)**

_ **Teaching Transparency, Life Cycle of an Angiosperm.** Use this transparency to discuss how the structure of teeth helps break down food. **(GENERAL)**

HOMEWORK

_ **Directed Reading Worksheet, Sexual Reproduction in Seed Plants, CRF.** Students complete the exercises in this worksheet to help them understand the material as they read the section. **(BASIC)**

_ **Active Reading Worksheet, Sexual Reproduction in Seed Plants, CRF.** Students read a passage related to the section topic and answer questions. **(GENERAL)**

Block 5

TEACH *(35 minutes)*

_ **Quick Lab,** Observing the Gametophytes of Pines, SE. Students observe the gametopytes of a pine with a microscope. **(GENERAL)**

_ **Datasheets for In-Text Labs, Observing the Gametophytes of Pines, CRF.**

_ **Teaching Transparency, Floral Structure.** Use this transparency to discuss how the structure of teeth helps break down food. **(GENERAL)**

_ **Quick Lab,** Observing the Arrangement of Parts of a Flower, SE. Students observe, compare, and contrast the parts of flowers. **(GENERAL)**

_ **Datasheets for In-Text Labs, Observing the Arrangement of Parts of a Flower, CRF.**

CLOSE *(5 minutes)*

_ **Quiz,** TE. Students answer questions that review the section material. **(GENERAL)**

HOMEWORK

_ **Reteaching,** TE. Students construct a crossword puzzle using the terms presented in this section. **(BASIC)**

_ **Teaching Tip**, Double Fertilization in Angiosperms, TE. Students make a graphic organizer to illustrate the process of double fertilization in angiosperms. A sample graphic organizer is provided in the TE. **(GENERAL)**

_ **Quiz, CRF.** This quiz consists of ten multiple choice and matching questions that review the section's main concepts. **(BASIC) Also in Spanish.**

_ **Section Review,** SE. Assign questions 1–6 for review, homework, or quiz. **(GENERAL)**

Optional Blocks

LABS *(90 minutes)*

_ **Exploration Lab, Fruits and Seeds, CRF.** Students observe and draw a variety of fruits and seeds. You will relate the form of some fruits and seeds to methods of dispersal. **(GENERAL)**

_ **Skills Practice Lab, Flower Structures, CRF.** Students examine a typical flower and observe the parts adapted for the production and protection of seeds. **(GENERAL)**

Other Resource Options

_ **Integrating Physics and Chemistry**, TE. Students measure the mass of pine seeds and time how long it takes each seed to reach the ground when dropped from different heights. Then students calculate each pine seed's acceleration using Newton's second law.

- **Alternative Assessment**, TE. Provide students with a variety of cones and fruits. Have students use a botany book to classify the fruits. **(GENERAL)**
- **Internet Connect.** Students can research Internet sources about Crop Pollination with SciLinks Code HX4053.
- **go.hrw.com.** For worksheets, videos, and other teaching aids related to this chapter, visit the HRW Web site and type in the keyword HX4 PRP.
- **CNN Science in the News, Video Segment 20 Migrant Bees.** This video segment is accompanied by a **Critical Thinking Worksheet**.
- **CNN Student News.** Find the latest news, lesson plans, and activities related to important scientific events at **cnnstudentnews.com**.

Lesson Plan

Section: Asexual Reproduction

Pacing

Regular Schedule:	**with lab(s):** 3 days	**without lab(s):** 2 days
Block Schedule:	**with lab(s):** 1 1/2 days	**without lab(s):** 1 day

Objectives

1. Describe several types of vegetative reproduction in plants.
2. Distinguish sexual reproduction in kalanchoës from asexual reproduction in kalanchoës.
3. Recommend several ways to propagate plants.

National Science Education Standards Covered

UNIFYING CONCEPTS AND PROCESSES

UCP1: Systems, order, and organization

UCP2: Evidence, models, and explanation

UCP5: Form and function

SCIENCE AS INQUIRY

SI1: Abilities necessary to do scientific inquiry

SI2: Understandings about scientific inquiry

LIFE SCIENCE: THE CELL

LSCell1: Cells have particular structures that underlie their functions.

LSCell4: Cell functions are regulated.

LSCell5: Plant cells contain chloroplasts, the site of photosynthesis.

LSCell6: Cells can differentiate and form complete multicellular organisms.

LIFE SCIENCE: INTERDEPENDENCE OF ORGANISMS

LSInter3: Organisms both cooperate and compete in ecosystems

KEY
SE = Student Edition TE = Teacher Edition
CRF = Chapter Resource File

Block 6

FOCUS *(5 minutes)*

_ **Bellringer Transparency.** Use this transparency as students enter the classroom and find their seats. **(GENERAL)**

MOTIVATE *(10 minutes)*

_ **Demonstration**, Observing Sweat, TE. Show students examples of bulbs, corms, rhizomes, runners, and tubers. Point out that all these are modified stems and therefore have buds that can grow into new plants. **(BASIC)**

TEACH *(30 minutes)*

_ **Teaching Transparency, Section Outline.** Use this transparency to give students a framework for the information in this section. **(GENERAL)**

_ **Teaching Transparency, Stems Modified for Vegetative Reproduction.** Use this transparency to discuss runners, bulbs, corms, rhizomes, and tubers. Encourage students to give additional examples of each types of modified stem. **(GENERAL)**

_ **Inclusion Strategies**, TE. Bring in a variety of vegetation for students to observe, touch, and sketch during class. **(GENERAL)**

_ **Up Close**, Kalanchoë, TE. Use the teaching strategies and discussion questions in the TE to guide students' reading of this illustrated feature in the SE. **(GENERAL)**

HOMEWORK

_ **Directed Reading Worksheet, Asexual Reproduction, CRF.** Students complete the exercises in this worksheet to help them understand the material as they read the section. **(BASIC)**

_ **Active Reading Worksheet, Asexual Reproduction, CRF.** Students read a passage related to the section topic and answer questions. **(GENERAL)**

Block 7

TEACH *(30 minutes)*

_ **Teaching Transparency, External Structures of Kalanchoë.** Use this transparency to describe the leaves, plantlets, flowers, air roots, and stem and leaf cuttings of a kalanchoë. **(GENERAL)**

_ **Teaching Transparency, Internal Structures of Kalanchoë.** Use this transparency to describe the leaf structure of a kalanchoë. and to discuss CAM photosynthesis in this plant's family. **(GENERAL)**

- **Teaching Transparency,** Methods of Vegetative Plant Propagation. Use this transparency to describe budding, grafting, cuttings, and tissue culture. **(GENERAL)**
- **Science Skills Worksheet, CRF.** Students complete a chart by classifying characteristics of plants as belonging to different plant groups. They also sequence the events in plant evolution. **(GENERAL)**

CLOSE *(15 minutes)*

- **Reteaching,** TE. Students describe the characteristics of plants that would indicate that they have been asexually propagated. **(BASIC)**
- **Quiz,** TE. Students answer questions that review the section material. **(GENERAL)**
- **Quiz, CRF.** This quiz consists of ten multiple choice and matching questions that review the section's main concepts. **(BASIC) Also in Spanish.**

HOMEWORK

- **Section Review,** SE. Assign questions 1–5 for review, homework, or quiz. **(GENERAL)**
- **Modified Worksheet, One-Stop Planner.** This worksheet has been specially modified to reach struggling students. **(BASIC)**
- **Critical Thinking Worksheet, CRF.** Students answer analogy-based questions that review the section's main concepts and vocabulary. **(ADVANCED)**

Optional Blocks

LAB *(45 minutes plus 10 minutes each day for 2 weeks)*

- **Exploration Lab,** Observing the Effects of Nutrients on Vegetative Reproduction, SE. Students investigate the effect of nutrients on the vegetative reproduction of duckweed. **(GENERAL)**
- **Datasheets for In-Text Labs, Observing the Effects of Nutrients on Vegetative Reproduction, CRF.**

Other Resource Options

- **Alternative Assessment,** TE. Students use poster board to design a garden that includes both sexually and asexually reproducing plants. **(GENERAL)**
- **Supplemental Reading, A Feeling for the Organism: The Life and Work of Barbara McClintock, One-Stop Planner.** Students read the book and answer questions. **(ADVANCED)**
- **Internet Connect.** Students can research Internet sources about Duckweed with SciLinks Code HX4062.
- **go.hrw.com.** For worksheets, videos, and other teaching aids related to this chapter, visit the HRW Web site and type in the keyword HX4 PRP.

_ **CNN Science in the News, Video Segment 20 Migrant Bees.** This video segment is accompanied by a **Critical Thinking Worksheet**.

_ **CNN Student News.** Find the latest news, lesson plans, and activities related to important scientific events at **cnnstudentnews.com**.

Lesson Plan

End-of-Chapter Review and Assessment

Pacing

Regular Schedule: 2 days

Block Schedule: 1 day

KEY	
SE = Student Edition	TE = Teacher Edition
CRF = Chapter Resource File	

Block 8

REVIEW *(45 minutes)*

_ **Study Zone,** SE. Use the Study Zone to review the Key Concepts and Key Terms of the chapter and prepare students for the Performance Zone questions. **(GENERAL)**

_ **Performance Zone,** SE. Assign questions to review the material for this chapter. Use the assignment guide to customize review for sections covered. **(GENERAL)**

_ **Teaching Transparency, Concept Mapping.** Use this transparency to review the concept map for this chapter. **(GENERAL)**

Block 9

ASSESSMENT *(45 minutes)*

_ **Chapter Test, Plant Reproduction, CRF.** This test contains 20 multiple choice and matching questions keyed to the chapter's objectives. **(GENERAL) Also in Spanish.**

_ **Chapter Test, Plant Reproduction, CRF.** This test contains 25 questions of various formats, each keyed to the chapter's objectives. **(ADVANCED)**

_ **Modified Chapter Test, One-Stop Planner.** This test has been specially modified to reach struggling students. **(BASIC)**

Other Resource Options

_ **Vocabulary Review Worksheet, CRF.** Use this worksheet to review the chapter vocabulary. **(GENERAL) Also in Spanish.**

_ **Test Prep Pretest, CRF.** Use this pretest to review the main content of the chapter. Each question is keyed to a section objective. **(GENERAL) Also in Spanish.**

_ **Test Item Listing for ExamView® Test Generator, CRF.** Use the Test Item Listing to identify questions to use in a customized homework, quiz, or test.

_ **ExamView® Test Generator, One-Stop Planner.** Create a customized homework, quiz, or test using the HRW Test Generator program.

TEST ITEM LISTING

Plant Reproduction

TRUE/FALSE

1. ____ Gametophytes produce zygotes and are diploid.
Answer: False Difficulty: I Section: 1 Objective: 1

2. ____ Mosses are found in all places and in all environments.
Answer: False Difficulty: I Section: 1 Objective: 1

3. ____ Spores form through mitosis in the spore capsule of mosses.
Answer: False Difficulty: I Section: 1 Objective: 1

4. ____ Water must be present for fertilization to take place in mosses.
Answer: True Difficulty: I Section: 1 Objective: 1

5. ____ Water must be present for fertilization to take place in ferns.
Answer: True Difficulty: I Section: 1 Objective: 2

6. ____ In ferns, *sorus* is another word for *frond.*
Answer: False Difficulty: I Section: 1 Objective: 2

7. ____ Haploid fern spores are produced by gametophytes.
Answer: False Difficulty: I Section: 1 Objective: 2

8. ____ Adult fern sporophytes are haploid.
Answer: False Difficulty: I Section: 1 Objective: 2

9. ____ In ferns, a large sporophyte with leaves called fronds alternates with a heart-shaped gametophyte.
Answer: True Difficulty: I Section: 1 Objective: 2

10. ____ Ferns need water to reproduce because their sperm must swim to eggs.
Answer: True Difficulty: I Section: 1 Objective: 2

11. ____ In ferns, the antheridia and archegonia are produced by different individuals.
Answer: False Difficulty: I Section: 1 Objective: 2

12. ____ Both mosses and ferns require a thin film of water for gametes to meet.
Answer: True Difficulty: I Section: 1 Objective: 3

13. ____ Unlike mosses, ferns show alternation of generations.
Answer: False Difficulty: I Section: 1 Objective: 3

14. ____ In both mosses and ferns, eggs are formed through mitosis in the antheridia.
Answer: False Difficulty: I Section: 1 Objective: 3

15. ____ In both mosses and ferns, sperm are formed through mitosis in the archegonia.
Answer: False Difficulty: I Section: 1 Objective: 3

16. ____ Mosses and ferns no longer require the presence of a film of water for reproduction.
Answer: False Difficulty: I Section: 1 Objective: 3

17. ____ Gametophytes of seed plants are large and easily viewed with the unaided eye.
Answer: False Difficulty: I Section: 2 Objective: 1

18. ____ In seed plants, gametophytes develop within the tissue of sporophytes.
Answer: True Difficulty: I Section: 2 Objective: 1

19. ____ Sexual reproduction in seed plants involves the transfer of eggs from the female reproductive structures of a plant to the male reproductive structures.
Answer: False Difficulty: I Section: 2 Objective: 1

20. ____ Pollination results from the fusion of male and female gametes.
Answer: False Difficulty: I Section: 2 Objective: 1

21. ____ The female gametophyte of a seed plant develops inside an ovule.
Answer: True Difficulty: I Section: 2 Objective: 1

22. ____ A plant embryo is a new sporophyte.
Answer: True Difficulty: I Section: 2 Objective: 2

23. ____ Seeds are a mechanism of survival for plants.
Answer: True Difficulty: I Section: 2 Objective: 2

24. ____ In flowering plants, the embryos of monocots have two cotyledons.
Answer: False Difficulty: I Section: 2 Objective: 2

25. ____ A mature pine tree produces either male or female cones.
Answer: False Difficulty: I Section: 2 Objective: 3

26. ____ Eggs and sperm are produced through mitosis in immature cones.
Answer: False Difficulty: I Section: 2 Objective: 3

27. ____ Gametophytes are physically much larger in conifers than their sporophytes.
Answer: False Difficulty: I Section: 2 Objective: 3

28. ____ In conifers, after pollination, a pollen tube grows from each pollen grain towards the eggs inside the ovule.
Answer: True Difficulty: I Section: 2 Objective: 3

29. ____ Many gymnosperms, such as pine, spruce, and fir trees, produce their seeds in cones.
Answer: True Difficulty: I Section: 2 Objective: 3

30. ____ Pollen is produced in the tip of the stamen, an area called the sepal.
Answer: False Difficulty: I Section: 2 Objective: 4

31. ____ In conifers, flower parts are arranged in four concentric whorls.
Answer: False Difficulty: I Section: 2 Objective: 4

32. ____ The purpose of petals is to attract pollinators such as birds and insects.
Answer: True Difficulty: I Section: 2 Objective: 4

33. ____ The lower portion of the pistil produces pollen.
Answer: False Difficulty: I Section: 2 Objective: 4

34. ____ Incomplete flowers consist of sepals, petals, stamens, and a pistil.
Answer: False Difficulty: I Section: 2 Objective: 4

35. ____ Gametophytes develop within flowers.
Answer: True Difficulty: I Section: 2 Objective: 4

36. ____ Virtually all of our food is derived, directly or indirectly, from flowering plants.
Answer: True Difficulty: I Section: 2 Objective: 4

37. ____ Pollen tubes grow through the style toward the ovule.
Answer: True Difficulty: I Section: 2 Objective: 5

38. ____ The flower of angiosperms are ornamental and have no reproductive function.
Answer: False Difficulty: I Section: 2 Objective: 5

39. ____ The fusion of a sperm with two nuclei in the ovule produces the endosperm of a seed.
Answer: True Difficulty: I Section: 2 Objective: 5

40. ____ Double fertilization has great survival value because each new generation carries its own initial source of nutrition.
Answer: True Difficulty: I Section: 2 Objective: 5

41. ____ All plants can reproduce only by sexual reproduction.
Answer: False Difficulty: I Section: 3 Objective: 1

42. ____ Asexual reproduction produces new plants that are genetically different from the parent plant.
Answer: False Difficulty: I Section: 3 Objective: 1

43. ____ Vegetative reproduction requires a plant to have flowers.
Answer: False Difficulty: I Section: 3 Objective: 1

44. ____ In vegetative reproduction, new plants may be generated from stems or roots.
Answer: True Difficulty: I Section: 3 Objective: 1

45. ____ Kalanchoë plants are capable of carrying out both sexual and asexual reproduction.
Answer: True Difficulty: I Section: 3 Objective: 2

46. ____ Kalanchoë plants produce adventitious roots.
Answer: True Difficulty: I Section: 3 Objective: 2

47. ____ Kalanchoë plants do not produce flowers.
Answer: False Difficulty: I Section: 3 Objective: 2

48. ____ Growing new plants from seeds or from vegetative parts is called plant propagation.
Answer: True Difficulty: I Section: 3 Objective: 3

MULTIPLE CHOICE

49. The structures on a moss gametophyte in which spores are produced are known as
a. meristems.
b. pollen grains.
c. cones.
d. sporangia.
Answer: D Difficulty: I Section: 1 Objective: 1

50. Which of the following statements about moss spores is *not* true?
a. They are produced by the sporophyte.
b. They are dispersed and then germinate.
c. They are produced in a capsule-like top.
d. They are diploid.
Answer: D Difficulty: I Section: 1 Objective: 1

51. The sporophyte generation in mosses produces spores by
a. meiosis.
b. mitosis.
c. sexual reproduction.
d. None of the above
Answer: A Difficulty: I Section: 1 Objective: 1

52. The dominant form of a moss life cycle is the
a. sporophyte.
b. gametophyte.
c. rhizoid.
d. zygote.
Answer: B Difficulty: I Section: 1 Objective: 1

53. In which of the following structures do liverworts and mosses produce eggs?
a. antheridia
b. capsules
c. archegonia
d. cones
Answer: C Difficulty: I Section: 1 Objective: 1

The diagram below shows the plant life cycle.

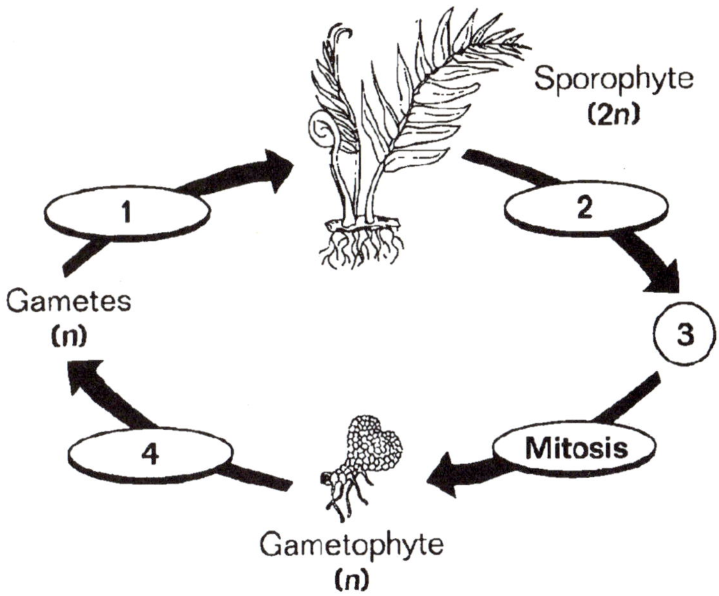

54. Refer to the diagram above. At which point in the life cycle does fertilization take place?
    a. 1
    b. 2
    c. 3
    d. 4

    Answer: A   Difficulty: II   Section: 1   Objective: 2

55. Refer to the diagram above. The structures produced at 3 are called
    a. sporangia.
    b. archegonia.
    c. spores.
    d. antheridia.

    Answer: C   Difficulty: II   Section: 1   Objective: 2

56. Fern sporophytes consist of rhizomes and
    a. flowers
    b. stems
    c. microgametophytes
    d. fronds

    Answer: D   Difficulty: I   Section: 1   Objective: 2

57. Ferns
    a. are found only in the tropics.
    b. require water for fertilization.
    c. produce diploid spores.
    d. have dominant gametophyte form.

    Answer: B   Difficulty: I   Section: 1   Objective: 2

58. Both mosses and ferns produce
    a. seeds.
    b. fruit.
    c. flowers.
    d. spores.

    Answer: D   Difficulty: I   Section: 1   Objective: 3

59. Unlike mosses, ferns possess
    a. spore capsules.
    b. archegonia.
    c. antheridia.
    d. vascular tissue.

    Answer: D   Difficulty: I   Section: 1   Objective: 3

60. Both mosses and ferns produce gametes through
    a. mitosis.
    b. meiosis.
    c. osmosis.
    d. diffusion.

    Answer: A   Difficulty: I   Section: 1   Objective: 3

61. antheridia : sperm ::
a. rhizoids : gametes
b. seeds : gametophytes
c. archegonia : eggs
d. sperm : megaspores

Answer: C Difficulty: II Section: 1 Objective: 3

62. In seed plants, female gametophytes produce
a. sperm.
b. eggs.
c. cells.
d. spores.

Answer: B Difficulty: I Section: 2 Objective: 1

63. A male gametophyte of a seed plant develops into a
a. pollen grain.
b. spore capsule.
c. pollen tube.
d. seed coat.

Answer: A Difficulty: I Section: 2 Objective: 1

64. A seed coat
a. protects the seed from injury.
b. prevents the embryo from dehydration.
c. protects the seed from disease.
d. All of the above

Answer: D Difficulty: I Section: 2 Objective: 2

65. Cotyledons
a. abosorb sunlight.
b. produce carbon dioxide.
c. store nutrients.
d. store water.

Answer: C Difficulty: I Section: 2 Objective: 2

66. The seeds of monocots have
a. one embryo and two cotyledons.
b. two embyos and two cotyledons.
c. one embryo and one cotyledon.
d. two embryos and one cotyledon.

Answer: C Difficulty: I Section: 2 Objective: 2

67. The cotyledons in a seed
a. protect the embryo.
b. provide a source of food for the embryo.
c. develop from the seed coat.
d. are part of the gametophyte.

Answer: B Difficulty: I Section: 2 Objective: 2

68. The partially developed plant found in seeds is known as a(n)
a. gametophyte.
b. spore capsule.
c. embryo.
d. sporophyte.

Answer: C Difficulty: I Section: 2 Objective: 2

69. A typical seed contains all of the following *except* a(n)
a. seed coat.
b. cotyledon.
c. embryo.
d. spore case.

Answer: D Difficulty: I Section: 2 Objective: 2

70. Which of the following were the first land plants to evolve seeds?
a. angiosperms
b. gymnosperms
c. mosses
d. ferns

Answer: B Difficulty: I Section: 2 Objective: 2

71. The process of transferring pollen from a male cone to a female cone in gymnosperms is called
a. fertilization.
b. seed formation.
c. pollination.
d. asexual reproduction.

Answer: C Difficulty: I Section: 2 Objective: 3

72. Pines, spruces, and firs are
    a. angiosperms.
    b. gymnosperms.
    c. flowering plants.
    d. sometimes nonvascular.

    Answer: B   Difficulty: I   Section: 2   Objective: 3

73. The tallest trees in the world are a species of
    a. conifers.
    b. dicots.
    c. liverworts.
    d. angiosperms.

    Answer: A   Difficulty: I   Section: 2   Objective: 3

74. In conifers, the diploid condition is resumed following
    a. pollination.
    b. fertilization.
    c. respiration.
    d. sporulation.

    Answer: B   Difficulty: II   Section: 2   Objective: 3

75. sporophytes : spores ::
    a. sporophytes : gametophytes
    b. gametophytes : gametes
    c. gametes : gametophytes
    d. sporophytes : gametes

    Answer: B   Difficulty: II   Section: 2   Objective: 3

76. Refer to the diagram above. The structure indicated at f
    a. supports the anther.
    b. produces pollen.
    c. supports the pistil.
    d. develops into a fruit.

    Answer: B   Difficulty: II   Section: 2   Objective: 4

77. Refer to the diagram above. The structure labeled c
    a. produces pollen.
    b. contains sperm cells.
    c. is the area where pollen lands and sticks.
    d. contains meristematic tissue.

    Answer: C   Difficulty: II   Section: 2   Objective: 4

78. Removing a flower's stigma would initially affect
    a. fertilization.
    b. seed production.
    c. pollination.
    d. seed dispersal.

    Answer: C   Difficulty: I   Section: 2   Objective: 4

79. Pollen is produced in a structure called the
    a. anther.
    b. stigma.
    c. ovary.
    d. pistil.

    Answer: A   Difficulty: I   Section: 2   Objective: 4

80. If a plant's flowers are very colorful and produce nectar, the plant is probably pollinated by
    a. water.
    b. wind.
    c. insects.
    d. self-pollination.

    Answer: C　Difficulty: I　Section: 2　Objective: 4

81. In angiosperms, immediately following pollination,
    a. the seed develops.
    b. an egg cell is formed.
    c. fertilization occurs.
    d. the pollen tube begins to form.

    Answer: D　Difficulty: I　Section: 2　Objective: 5

82. In angiosperms, fertilization
    a. involves the union of the egg and sperm.
    b. may not follow pollination at all.
    c. may not occur until weeks or months after pollination has taken place.
    d. All of the above

    Answer: D　Difficulty: I　Section: 2　Objective: 5

83. During fertilization in flowering plants, one sperm fuses with an egg to form an embryo, and another fuses with two nuclei to form nutritive tissue. This event is called
    a. self-pollination.
    b. adaptation.
    c. maximization.
    d. double fertilization.

    Answer: D　Difficulty: I　Section: 2　Objective: 5

84. Angiosperms produce spores that grow into gametophytes through the process of
    a. mitosis.
    b. meiosis.
    c. osmosis.
    d. All of the above

    Answer: B　Difficulty: I　Section: 2　Objective: 5

85. In angiosperms, flowers are
    a. photosynthetic.
    b. autotrophic.
    c. sporophytes.
    d. gametophytes.

    Answer: C　Difficulty: I　Section: 2　Objective: 5

86. Which structure allows plants to reproduce asexually?
    a. rhizome
    b. flower
    c. archegonium
    d. antheridium

    Answer: A　Difficulty: II　Section: 3　Objective: 1

87. The production of offspring genetically identical to the parent plant is the result of
    a. sexual reproduction.
    b. asexual reproduction.
    c. alternation of generations.
    d. double fertilization.

    Answer: B　Difficulty: I　Section: 3　Objective: 1

88. Plantlets are
    a. tiny, new plants that develop at leaf margins.
    b. a means of vegetative reproduction in kalanchoe.
    c. genetically identical to the parent plant.
    d. All of the above

    Answer: D　Difficulty: I　Section: 3　Objective: 2

89. Kalanchoes are succulents, meaning that they
    a. have fleshy leaves and stems to store water.
    b. have a large taproot that stores water.
    c. require a large amount of water from the soil.
    d. have stomata that open only in the daytime.

    Answer: A　Difficulty: I　Section: 3　Objective: 2

90. Growing new plants from seed or from vegetative parts is called
    a. sexual reproduction.
    b. double fertilization.
    c. plant propagation.
    d. binary fission.

    Answer: C   Difficulty: I   Section: 3   Objective: 3

COMPLETION

91. Mosses have a life cycle called ____________ ____________ ____________.

    Answer: alternation of generations   Difficulty: I   Section: 1   Objective: 1

92. The tiny liverwort ____________ grow from the archegonia under the caps of female stalks.

    Answer: sporophytes   Difficulty: I   Section: 1   Objective: 1

93. The gametophytes of ____________ plants are larger and more noticeable than the sporophytes.

    Answer: nonvascular   Difficulty: I   Section: 1   Objective: 1

94. In ferns, the diploid sporophytes produce spores by ____________.

    Answer: meiosis   Difficulty: I   Section: 1   Objective: 2

95. Like the nonvascular plants, the seedless vascular plants can only reproduce sexually when a film of ____________ covers the gametophyte.

    Answer: water   Difficulty: I   Section: 1   Objective: 3

96. Unlike mosses, ferns have sporophytes that are much ____________ than their gametophytes.

    Answer: larger   Difficulty: I   Section: 1   Objective: 3

97. Both mosses and ferns produce eggs within structures called ____________.

    Answer: archegonia   Difficulty: I   Section: 1   Objective: 3

98. Both mosses and ferns produce sperm within the ____________.

    Answer: antheridia   Difficulty: I   Section: 1   Objective: 3

99. A male gametophyte of a seed plant develops into a(n) ____________ ____________.

    Answer: pollen grain   Difficulty: I   Section: 2   Objective: 1

100. A(n) ____________ is a multicellular structure that is part of the sporophyte.

    Answer: ovule   Difficulty: I   Section: 2   Objective: 1

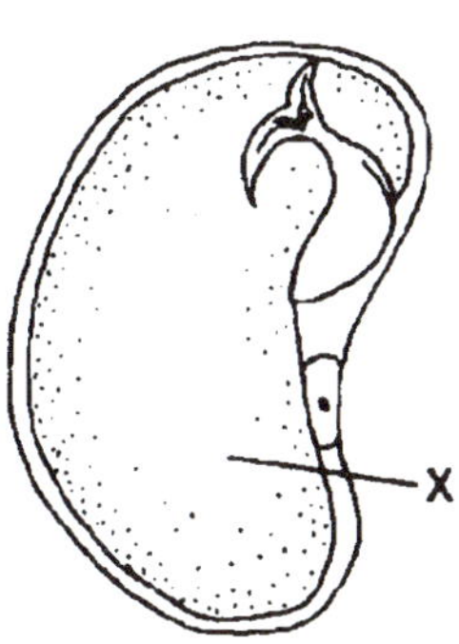

101. Refer to the diagram above. The outer structure in the diagram above is called the ____________ ____________.

    Answer: seed coat   Difficulty: II   Section: 2   Objective: 2

102. Refer to the diagram above. The structure labeled X is called the ____________________.
Answer: cotyledon Difficulty: II Section: 2 Objective: 2

103. A mature pine tree produces both male and female ____________________.
Answer: cones Difficulty: I Section: 2 Objective: 3

104. During ____________________, a haploid sperm fuses with a haploid egg, forming a diploid zygote.
Answer: fertilization Difficulty: I Section: 2 Objective: 3

105. In conifers, a very large sporophyte that produces cones alternates with tiny gametophytes that form on the ____________________ of cones.
Answer: scales Difficulty: I Section: 2 Objective: 3

106. Refer to the diagram above. The structure labeled c is called the ____________________.
Answer: stigma Difficulty: II Section: 2 Objective: 4

107. Refer to the diagram above. The structure labeled f is called the ____________________.
Answer: anther Difficulty: II Section: 2 Objective: 4

108. Refer to the diagram above. The structure labeled b is called the ____________________.
Answer: sepal Difficulty: II Section: 2 Objective: 4

109. Refer to the diagram above. A flower like this one has all four basic flower parts and thus is an example of a(n) ____________________ flower.
Answer: complete Difficulty: II Section: 2 Objective: 4

110. Flowers are a source of ____________________ for pollinators.
Answer: food Difficulty: I Section: 2 Objective: 4

111. The transfer of pollen grains from an anther to a stigma is known as ____________________.
Answer: pollination Difficulty: I Section: 2 Objective: 5

112. In angiosperms, seeds develop from the ____________________ after an egg has been fertilized.
Answer: ovules Difficulty: I Section: 2 Objective: 5

113. The seeds of angiosperms are enclosed in ____________________.
Answer: fruits Difficulty: I Section: 2 Objective: 5

114. The event in which one sperm fertilizes an egg and a second sperm fuses with two nuclei is called ____________________ ____________________.
Answer: double fertilization Difficulty: I Section: 2 Objective: 5

115. Many of the structures by which plants reproduce vegetatively are modified ____________________.
Answer: stems Difficulty: I Section: 3 Objective: 1

116. Bulbs, corms, rhizomes, and tubers are examples of modified stems that allow plants to reproduce ____________________.

Answer: vegetatively Difficulty: I Section: 3 Objective: 1

117. Kalanchoes and other succulents that fix carbon by a water-conserving pathway are called ____________________ plants.

Answer: CAM (crassulacean acid metabolism)

Difficulty: I Section: 3 Objective: 2

118. The technique called ____________________ ____________________ involves taking a small piece of tissue from a plant and placing it on a nutrient-rich medium to produce large numbers of identical plants.

Answer: tissue culture Difficulty: I Section: 3 Objective: 3

119. The process known as ____________________ involves attaching small stems from one plant to larger stems or roots of another plant.

Answer: grafting Difficulty: I Section: 3 Objective: 3

ESSAY

120. Compare and contrast nonvascular plants with seedless vascular plants.

Answer:

Eggs are produced in archegonia, and sperm are produced in antheridia in both nonvascular and seedless vascular plants. Both plant types require the presence of a thin film of water to enable sperm to come into contact with eggs. Unlike vascular plants, nonvascular plants lack organized tissues that move water and nutrients a great distance within the plant body. Additionally, the sporophyte is significantly larger than the gametophyte in vascular plants. This condition is reversed in nonvascular plants.

Difficulty: III Section: 1 Objective: 3

121. Describe the function of the seed coat.

Answer:

The outside of the seed is covered by a protective seed coat. The seed coat prevents the embryo from drying out and protects the embryo from mechanical injury and disease.

Difficulty: II Section: 2 Objective: 2

122. Outline the processes of pollination and fertilization in gymnosperms.

Answer:

Pollen produced by pollen cones are carried by the wind to female cones that have opened, exposing the ovules. After pollen grains land near ovules, slender pollen tubes grow out of the pollen grains and into the ovules. Sperm move through the pollen tubes and enter the ovules. Once inside the ovules, fertilization occurs when sperm fuse with eggs, forming zygotes.

Difficulty: III Section: 2 Objective: 3

123. List and describe the location of the parts of a complete flower.

Answer:

A complete flower is one that contains all four floral parts. The outermost whorl of the flower is made up of sepals. Proceeding toward the interior, the second whorl is made up of petals. The third whorl is made up of stamens. At the center of the flower is the pistil. Dependent upon the species, a flower may possess more than one pistil.

Difficulty: II Section: 2 Objective: 4

124. Describe a unique method of vegetative reproduction in kalanchoë.

Answer:

Tiny, new plants called plantlets develop in notches along the leaf margins, fall to the ground, and then grow into new plants.

Difficulty: II Section: 3 Objective: 3

9997266021 1 2 3 4 5 6